DATE DUE			
GAYLORD			PRINTED IN U.S.A.

Product Concept Design

Turkka Keinonen and Roope Takala (Eds.)

Product Concept Design

A Review of the Conceptual Design of Products in Industry

With 70 Figures

 Springer

Turkka Keinonen, PhD
University of Art and Design Helsinki
Hämeentie 135 C
FIN-00560 Helsinki
Finland

Roope Takala, MSc
P. O. Box 407
FIN-00045 Nokia Group
Finland

British Library Cataloguing in Publication Data
Product concept design: a review of the conceptual design
   of products in industry
   1. New products - Management
   I. Keinonen, Turkka II. Takala, Roope
   658.5'752
ISBN-10: 1846281253

Library of Congress Control Number: 2005935663

ISBN-10: 1-84628-125-3   e-ISBN 1-84628-126-1   Printed on acid-free paper
ISBN-13: 978-1-84628-125-9

Printed in Germany

9 8 7 6 5 4 3 2 1

Springer Science+Business Media
springer.com

Preface

Concepting is a relatively new idea within product development. However, product design veterans would undoubtedly say that this type of activity has existed for a long time. Nevertheless, the topical nature of concepting can be understood by taking a brief look at the history of innovation processes (Figure 1)[1,6].

The first-generation innovation processes introduced in the 1950s and 1960s were based around the technological capabilities of the time. Huge leaps in technological development were constantly pushing new products onto the market without leaving time to consider the user's requirements. During the 1970s, production continued to increase and companies struggled to become global businesses, which encouraged them to look at technology and their products from a volume market perspective. As a result, the second-generation innovation processes can be seen as market-driven activities. The aim of innovation was to meet market needs by introducing further technological developments.

The oil crisis of the 1970s and the resulting problems with the world economy forced companies to rationalise their operations. They also began to regard innovation processes more as an entity in themselves and as a part of the efficient operation of the whole company. This approach prevailed until the middle of the 1980s. At the same time, the available resources were becoming scarce and it was no longer enough simply to improve the quality of products. The third-generation innovation processes, for the first time, systematically linked the technology, the markets and the functions of the organisation. The aim was to take the needs of the markets and the opportunities afforded by the technology into consideration by carrying out thorough research and testing, and by making sure of the acceptance of new products, so that they genuinely met market needs. This kind of activity required companies to be methodical, to have good internal and external communications and to be able to learn from each process. Innovation was perceived as a process that linked the organisation's internal abilities to the external operators and markets.

In the 1990s, the innovation focus of companies shifted to generic technologies, and the evolving information technology (IT) sector redirected the technology strategy towards developments in production and manufacturing. At the same time, companies focused on their core competencies, and the globalisation of the markets forced them to enter into alliances and collaborative networks. This, and the development of the IT sector, led to a dramatic shortening in the product life cycle, which in turn forced companies to boost their operational efficiency further because the first to market gained the competitive edge. The networking of companies, concurrent design, production and marketing made short product design cycles a reality. Fundamental to the fourth-generation innovation processes was continuous cooperation with the markets. Customers were closely linked to product development, and the environment was systematically mapped in order to acquire new information. The product development process consisted of a group of overlapping functions. Markets were constantly probed, and product design and production took place concurrently. The innovation process was integrated into overlapping functions. Design was no longer an activity that preceded production and marketing; instead it was carried out in parallel. The links between design, production and marketing became significantly more complex than in the previous generations.

The third generation highlighted the importance of the organisation's anticipatory external information for the different innovation process generations. Systematic research into markets began, and not only the trends but also the forthcoming needs were anticipated, for example, by drafting alternative future outlooks and by designing different technology strategies to accommodate them. In the fourth generation, the speed of product development became sufficient to allow the anticipation of market behaviour from weak signals. Therefore the development of the environment must be tracked continuously and future needs must be anticipated as early and accurately as possible. Only the third and, in particular, the

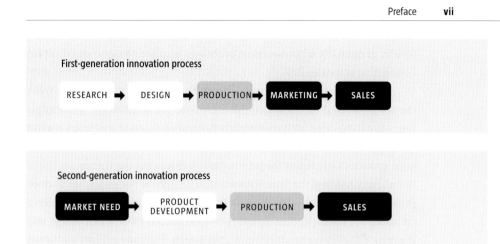

First-generation innovation process

RESEARCH → DESIGN → PRODUCTION → MARKETING → SALES

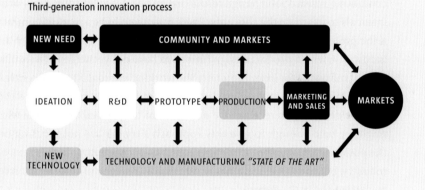

Second-generation innovation process

MARKET NEED → PRODUCT DEVELOPMENT → PRODUCTION → SALES

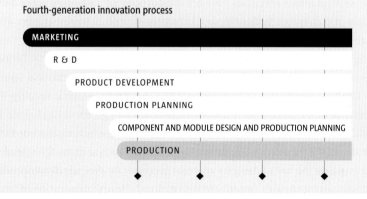

Third-generation innovation process

NEW NEED ↔ COMMUNITY AND MARKETS

IDEATION ↔ R&D ↔ PROTOTYPE ↔ PRODUCTION ↔ MARKETING AND SALES ↔ MARKETS

NEW TECHNOLOGY ↔ TECHNOLOGY AND MANUFACTURING *"STATE OF THE ART"*

Fourth-generation innovation process

MARKETING

R & D

PRODUCT DEVELOPMENT

PRODUCTION PLANNING

COMPONENT AND MODULE DESIGN AND PRODUCTION PLANNING

PRODUCTION

FIGURE 0.1.
Rothwell's four generations of innovation processes.[1]

fourth generation of innovation processes require companies to set up concepting activities.[1,6]

The concept design case studies in this book are taken mainly from the transport and electronics industries. Both are technology- and design-intensive industries with major players competing on global markets and making huge investments in product development. They are industries that need proactive strategies for probing and influencing the future, and have definitively reached the fourth generation. In light of the examples that we present, it would seem that the best conditions for concept design within a company are when product design processes already exist, when design and the responsibility for production-related design is already in place and when the new-product development activities (NPD) have matured. Under these conditions, research and design resources can be released continuously or at intervals to examine the concepts. There may seem to be a marked contrast if the projects and business models that we describe are compared with an industry in which product design functions have not yet achieved a similar level of maturity, for example where the working approach to user-centred design, innovation activities or design has not been incorporated into the product design. Nevertheless, we do not want to frighten off anyone who might be bold enough to give this approach a try. We are not focusing on the possibilities, significance or prerequisites of concepting in this type of industry, but there will certainly be a need for this type of investigation to be carried out in the future.

This book is based on an earlier volume published in Finnish[2]. However, several major revisions have been incorporated, including some completely new chapters. Both of the books have been written by researchers who represent both academic institutions and business: the University of Art and Design Helsinki, the Helsinki University of Technology, the Helsinki School of Economics, Nokia and the design consultancies Muotohiomo and Sweco Pic. The National Technology Agency of Finland, Tekes, which is running an industrial design technology programme called Design 2005, has helped to facilitate the collaborative process of writing and researching the books.[3] The aim of the programme has been to raise the standard of design research and improve the use of design expertise in corporate product development and business strategies, in order to move into the fourth generation of innovation. The programme forms part of the Finnish government's resolution on design policy dated June 15, 2000[4]. The last chapter of the book also includes contributions related to work carried out as part of Masina, another Tekes programme which aims to improve product design in the machine design industry[5]. Even though the programmes are Finnish and all the contributors are Finns, we have aimed to give the book an international perspective. Our material includes several case studies from companies operating in other European countries and on the global market.

The book is aimed at professionals, students of design, design management, product development and product marketing, and at anyone who is interested in learning about design in a wider business context and about approaches to innovative NPD. For this reason the book is written in an approachable style. However, the book is based in part on original research, conference papers, journal articles and academic theses written by the contributors. References are made to these in each chapter. Some chapters are based on secondary sources and the contributors' first-hand experience in industry. We hope that these different angles on concepting make the book interesting and give it a broad scope, without confusing the reader.

We begin in Chapter 1 with a discussion on the differences between product and concept design by defining concepting and presenting the generic objectives of concept design. We demonstrate that concept design when applied as a strategic tool has a very broad interface to several corporate functions in addition to product development. The second chapter gives a player's perspective on concept creation by highlighting issues related to structuring and facilitating the work of concept design teams. Chapter 3 describes concept design activities from the point of view of the process, including, for example, the assessment and road mapping of concepts. The key issue in planning a concepting process is to enable creative exploration whilst simultaneously managing the work involved. User information and user-centred design are regarded as crucial foundations for creating concepts that are meaningful to customers and commercially profitable. Chapter 4 discusses the concept of user experience and addresses the different relationships and design approaches that design teams and user communities may have.

Chapters 5 to 7 put concept creation into different business contexts. Chapter 5 describes concept design activities in a mature industry, in this case the automotive industry, where radical concepts are seldom introduced as products, but where the companies experiment with changes and design variations within their concepts. Chapter 6 questions the distinction between products and concepts by arguing that in fast-moving and dynamic businesses, such as communication technology, the manufacturers know so little about the markets of the future that they have to implement and launch concepts to learn about the customers' preferences and behaviour. The last chapter (Chapter 7) introduces a concept design process that is applicable when companies want to learn about their opportunities in the future, which is beyond the normal scope of product development. Their current customers and competitors do not significantly influence the concepting process. Instead, the vision concepts are based on future research and far-reaching future scenarios.

The editors would like to thank all the contributors to this book, the contributors to the previous Finnish book, the company representatives who agreed to be interviewed and everyone else whose help and cooperation made the completion of this work possible.

Helsinki, June 24, 2005
Turkka Keinonen, Roope Takala

References

1   **Rothwell, R.** (1994): Towards the fifth-generation innovation process. International Marketing Review vol. 11, issue 7, pp. 7-31

2   **Keinonen, T. and Jääskö V. eds.** (2003): Tuotekonseptointi, [Product Concepting] Teknologiainfo Teknova, Helsinki.

3   **Design 2005 Technology Programme** http://websrv2.tekes.fi/opencms/opencms/ OhjelmaPortaali/Kaynnissa/MUOTO_2005/en/etusivu.html (June 24, 2005)

4   **Government Decision in Principle on Design Policy**, June 15, 2000 http://www.minedu.fi/ julkaisut/vnmuotoilu.html (August 21, 2003)

5   **Masina Technology Programme** http://websrv2.tekes.fi/opencms/opencms/OhjelmaPortaali/ Kaynnissa/MASINA/en/etusivu.html (June 24, 2005)

6   **Keinonen, T., Andersson, J., Bergman, J.-P, Piira, S. & Sääskilahti, M.** (2004): Mitä tuotekonseptointi on? [What is Product Concepting?] in Keinonen, T. & Jääskö, V. eds. Tuote-konseptointi. Teknologiainfo Teknova, Helsinki.

Contents

1

Introduction to Concept Design

Turkka Keinonen

1   Introduction to Concept Design[1]

Turkka Keinonen

1.1   About design, delivery and manufacturing

Product design is customarily linked to manufacturing; products fulfil the needs of customers, and business is built on the exchange of the products. Engineering supports production, which in turn provides goods and services to promote a prosperous society. Product designers draft technical drawings, models and component lists specifying the shape of product components and the ways in which parts are interconnected, so that workshops and factories can produce material and software commodities. The demands of manufacturing define the design process by separating the possible and the impossible solutions, and classifying the feasible solutions as either economically viable or not cost-effective. Production dictates the level of precision and the scope of the specifications expected from the design process. Some of the results of the product design process do not reach the production stage, whereas others function as tools that support the design process and promote understanding and discussion during the design phase. Even then,

---

1 This chapter is an edited version of an original text[1] published in Finnish. We acknowledge the contributors to the original text: Janne Andersson, Jukka-Pekka Bergman, Sampsa Piira and Mikko Sääskilahti.

the goal of product design is to refine the initial outlines into specifications that will guide production.

Production is connected to distribution. The purpose of distribution is to bring the goods to the marketplace so that consumers become aware of them and have access to them. The timing of the market launch, the number of products to be supplied, the quality characteristics are, in turn, the requirements that define the production process. They are reflected through production as design requirements. So, as a subordinate function to production and distribution, product design must fulfil several requirements from these processes, including the degree of detail in the specifications, the internal accuracy of the specifications, the compatibility with production and the accurate timing of the specification delivery. Each component of the product must be specified down to the last detail so that the components fit together and so that its production is economically viable.

In order for product design to fulfil the requirements mentioned, several methods for supporting design activities have been developed. Quality systems and certifications ensure the intercompatibility of the design's starting points and end results, concurrent engineering helps to squeeze as much design work as possible into as short a time as possible and computer-aided design systems support the precise and detailed specification of the product's geometry. The use of design tools facilitates the creation of specifications suitable for production and, at the same time, characterises product design as a function. Product design is a function that focuses systematically on the

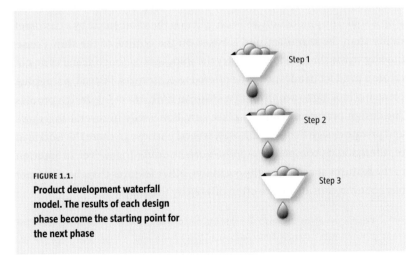

Step 1

Step 2

FIGURE 1.1.
Product development waterfall model. The results of each design phase become the starting point for the next phase

Step 3

development of precise and unambiguous specifications, and has scheduled performance requirements. The focus on one solution is characteristic of the waterfall (or Stage-Gate™ [2]) models for design work (Figure 1.1). The process advances in one direction, and only one solution at a time can fit through the small opening of the output funnel.

Sometimes, however, and lately with increasing frequency, product designers work on projects in which production and distribution do not directly dictate the environment and requirements of the design. Judging by the practices used, these projects have a clear connection with the world of product design, but they do not seem to meet many of the fundamental product design characteristics described above. As the following KMY Technology, Ed-design and Whirlpool case studies (and others in later chapters) show, this type of activity results in designs that do not specify all the product details, omit some of the requirements set by marketing and, as such, are technically unfeasible. However, projects realised in this way are being given a high public profile rather than the quiet burial that is the common fate of failed designs. Let us first look at some of these cases more closely and then come back to discuss their objectives on a more generic level.

1.2   Visionary cruise liners

The difference between shipbuilding and many other industries that involve delivering finished goods to the marketplace for a customer to choose from is that the final design and construction of a ship does not start until the ship is sold. The production decision, which is therefore made by a customer rather than the manufacturer, is based on the outline of the ship. Consequently, in the shipbuilding industry the shipyards or design agencies do not create a product in advance to be offered to customers. Instead, a shipping company looking to acquire a new ship can itself direct the design process. The design process often takes place simultaneously in several shipyards and private design firms. In this way, several parties can share the workload or, alternatively, competing proposals can be drafted. However, in addition to the features specified by the customer, other features that the customer has not even considered are often offered by the supplier.

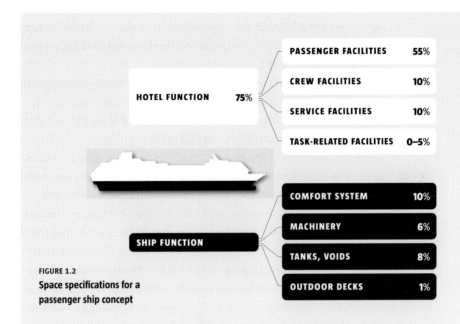

FIGURE 1.2
Space specifications for a passenger ship concept

Kvaerner Masa-Yards Technology (KMY Technology)[2] is a division of Kvaerner Masa-Yards Inc. that provides R&D, concept design, shipyard technology and technology transfer services. KMY Technology's ship technology unit focuses on ship-related R&D and design. According to industrial designer Janne Andersson, the initial phase of KMY Technology's product planning process relies on function-based ship design, which defines the vessel's mission and operation plan. After that, the vessel's functions and central systems are described and used to outline space specifications. Space specifications include, for example, the deck area and other capacity needs (Figure 1.2). Using the data from the space specifications and the statistics collected within the company, the preliminary weight and cost of the vessel are calculated. These figures are then used to determine the vessel's main dimensions, performance and economic efficiency. After a rough general layout has been produced, the ship design can take on a more detailed focus. Initially, the accuracy and scope of the design solutions presented for reviews

---

2 Since this study was carried out, Kvaerner Masa-Yards and Aker Finnyards have merged and the new company uses the name Aker Finnyards. Here we have used the name KMY Technology, as the information in this chapter is based on the activities of this business unit.

and comments are such that they can be updated quickly and with relatively little effort. The content changes as the development work progresses, becoming more detailed as it nears completion.

KMY Technology draws up product proposals for three groups among the technology unit's clientele: external customers who are typically shipping companies, the parent company's own shipyards and suppliers of ship systems, such as engine manufacturers and cabin suppliers.

The designs created for external customers vary widely in content and scope. The most common type of project is based on the customer's specifications. More imaginative and visionary trials are carried out, for instance, if the customer is considering a business idea that ventures beyond conventional seafaring or if the company ordering the ship belongs to a different sector of industry, but is planning to become a shipowner. In this type of case, redefining the cruise liner as we know it today would not meet the requirements of the new business models. System suppliers sometimes commission an entire ship outline to showcase the advantages that their new product offer. The design might highlight the space savings that can be achieved in the general facilities, for instance, or the system's effect on the positioning of other equipment. Figure 1.3 shows a design ordered by an engine manufacturer, which functions as a presentation platform for the company's new space-saving engine concept.

Some of the designs produced for external customers and system suppliers involve testing the potential and functionality of a new technology or business model. In these cases the design is investigative in nature, and the time span of the design might be longer than usual. When completed, some of these designs are put on hold to await a more opportune time, and many of them are never intended as a preparatory stage in the construction of ship – they are instead produced purely as studies.

KMY Technology also designs visionary future ships without commissions from external customers. According to Andersson, futuristic designs are ideally suited to marketing the image of the company because they convey the impression of a company that is ahead of its time. The driving forces behind these designs are often new technological developments and the renewal of sectors of the business. A method that is frequently used to anticipate future developments is trend extrapolation, in which shipbuilding trends are tracked using comprehensive statistical data relating to vessel dimensions such as the deck area, power outputs ship-related expenses and

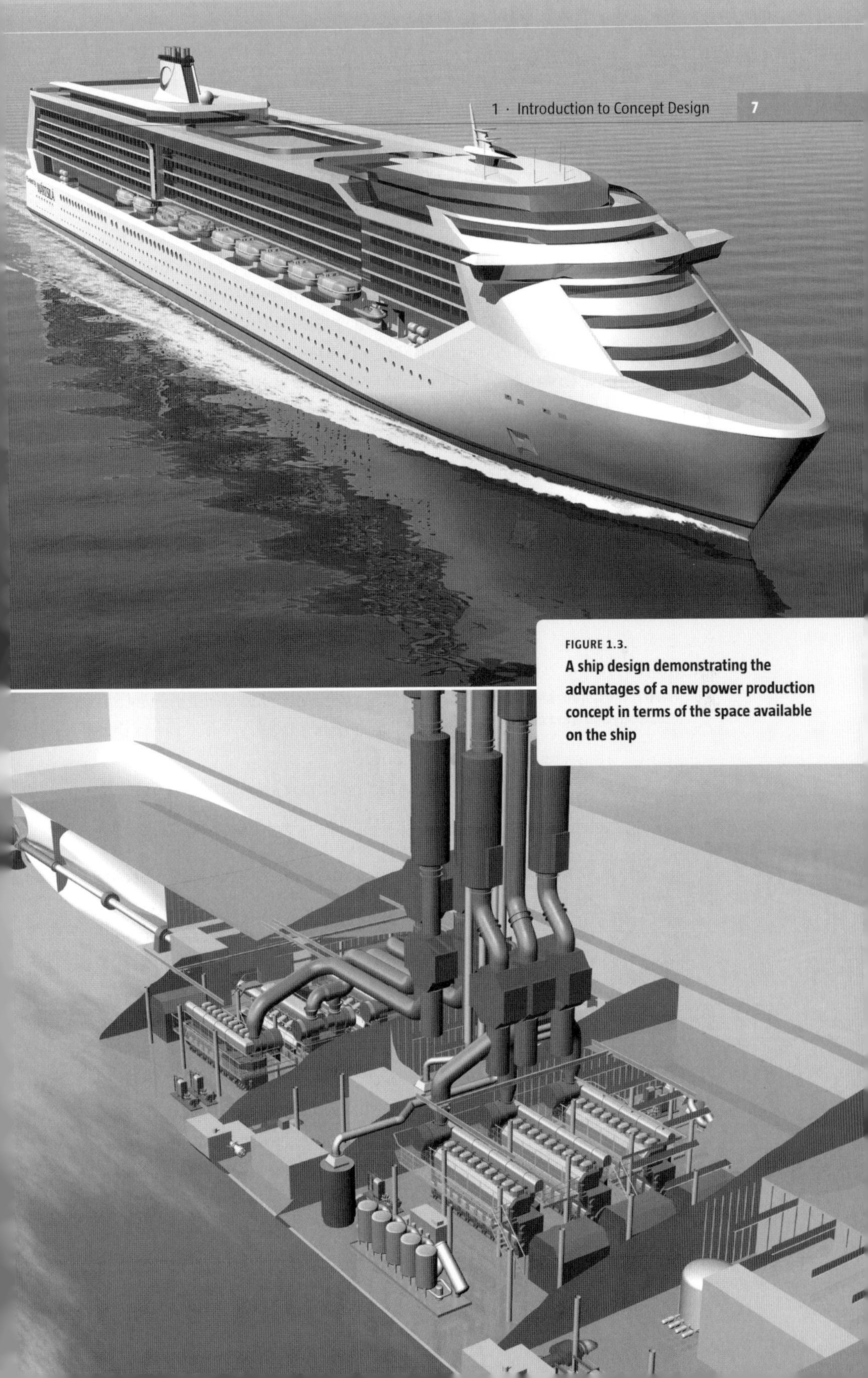

FIGURE 1.3.
A ship design demonstrating the advantages of a new power production concept in terms of the space available on the ship

revenue. The different variables of a ship being designed can be compared with these statistical data, or a statistically based trend extrapolation can be used as the starting point for a new ship. For instance, statistical data show that there is a link between the size and profitability of a cruise ship: increasing the size of a ship results in economies of scale. From the passengers' perspective a ship functions as a holiday destination, with larger ship offering a wider variety of services and activities. In addition, the proportion of cabins with windows is increasing in a larger ship, which has a major impact on the architecture of the ship.[1]

Another starting point for future designs is to anticipate the growth of shipping markets and new routes opening up. One such example is the route between the USA and Cuba. Ship designs have already been developed in anticipation of the opening of this route. Another is the New York–Bahamas route, which may become profitable as a result of the increased speed of the ships being developed for the future.

FIGURE 1.4.

A ship concept from KMY Technology, based on fuel cell technology and wind energy

In addition to the tracking of trends and the systematic collection of data, the experience and insight of professional designers also contribute to the ship designs of the future. For instance, Figure 1.4 shows a futuristic, vision-based concept involving a more environmentally friendly ship based on fuel cell technology and wind energy. Up to ten people, including engineers, designers and architects, typically participate in the design process. The amount of creative work involved in informal development studies of future cruise ships is significant. Visionary ship designs are used to attract customers and even to generate new needs. Examples of this include on-board water parks and promenades, the possibilities of which customers may not initially be aware of. According to Andersson, another of the added values of futuristic design is the increase in innovativeness and creativity amongst employees. This also helps to promote the entire industry, because by bringing the ship designs to the attention of the public, the industry can demonstrate the existence of new, untapped possibilities.

Ship technology evolves because technical developments have to respond to the challenges set by the designs. Integrating the design-driven promenade solution – which is an open walking area running the entire length of the ship – into the ship's structure was still a major technical challenge only a few years ago. Development also takes place in the reverse direction, where technological innovation creates new design possibilities. For example, it would have been very difficult – if not impossible – to build the huge Voyager class of Caribbean cruise ships without the invention of the Azipod propulsion system.

1.3   The future of home, work and leisure

Ed-design, a company based in Turku, Finland, is one of the biggest industrial design firms in the Nordic countries. Every Ed-design project initially aims to look into the future, typically pursuing a vision of 2 to 5 years ahead. According to traditional design practice, the way to anticipate the future is to evaluate technological and market trends on the basis of the designers' personal knowledge and long-term experience. The restrictions and the design goals for making the future products the "most advanced, yet acceptable", as Raymond Loewy has put it, are then put in place, taking into account the trends that have been identified.

In 2000, as part of Ed-design's 10th anniversary, the firm decided to focus on understanding the future. According to managing director Tapani Hyvönen, one specific future-oriented project was intended to generate the background material for forthcoming customer projects. In a project that aimed far beyond the normal scope, it was also possible to break away from ordinary design routines and to motivate employees to do something different.

The project started by collecting information about technology and product trends from the 1970s to 2000s. The information was compiled into a matrix on the wall of the managing director's office, with the location intended to provide a clear indication of management's commitment to the project. The trends identified were extrapolated as far into the future as possible, based on the designers' knowledge and insight. The resulting trend graphs were examined from the perspective of the different focuses of the project – home, work and leisure. Having identified the main trends, the next task was to develop scenarios and product ideas around them. In the next phase video scripts were written using scenarios (Figure 1.5). Some of the product ideas were elaborated into mock-ups and used as props in the videos. A video production company was brought in to produce the films and took responsibility for casting, set design, video technology and directing, with Ed-design's designers supervising the realisation of their ideas.

For Ed-design, the work process as such was almost as important as the final result. Hyvönen noted how important it is for a design agency to be continuously visualising the future, because the information produced by

FIGURE 1.5.

Ed-design's future scenario of technologies and services for home, work and leisure

this process reduces the design risks in forthcoming projects. The images of the future produced in the video project were used in marketing to demonstrate the company's innovativeness. However, the most beneficial aspects of the process were the opportunities for learning and for improving employees' motivation. The project helped to gather and analyse ideas for joint use and to develop a team spirit within the company.

1.4   The biological laundry

As the world's leading manufacturer and marketer of home appliances, Whirlpool, is famous for its refrigerators, washing machines and other white goods. The company has production sites in 13 countries and markets its 11 largest brand names in more than 170 countries.

According to Chuck Jones, vice president of the global consumer design division (GCD), design touches three primary domains at Whirlpool: brand, product and experience. With regard to brand, design influences the development of brand architecture, essentially working with the brand to define what it is, who it reaches and how. For the product dimension, design works along strategic and tactical paths. For strategic purposes GCD periodically develops design exploration exercises such as Project F (described below) which are a way for Whirlpool teams to stretch and dream. For tactical purposes, a visual brand language is followed, which is essentially a toolkit to allow the designers to carefully manage the visual brand equity of a product. The last area, experience, is primarily driven through Whirlpool's global usability organization and is tightly linked to developing shaped consumer experiences.

Whirlpool has adopted a user-centred design approach, which shifts the focus of design from how the product looks to how the product is used. This is assured by having industrial design and usability linked together in a common organization. Whirlpool's design-driven research projects strive first and foremost to understand the user and the context where the products are used, and consumer understanding is increasingly shaping which technologies are explored by a company.

Jones believes that designers are exceptionally good at creating compelling visions. According to him, design is uniquely positioned to create visualizations and scenarios to draw other parts of the organisation in and give them something to rally around and support. Jones' analogy is building a constituency, that is to understand who the constituents are and leveraging all design tools such as mock-ups, renderings, animations so as to create a vision that is so compelling that people want to become involved and support it. The scenarios created by designers combine various pieces of information and take the research forward. Bringing the information together in a format that is easy to understand helps to uncover new possibilities and anticipate future changes.

Project F is presented here as an example of Whirlpool's research and concepting projects. The purpose of Project F was to increase the company's knowledge and understanding of clothing care in the future and its effects on the manufacturing of domestic appliances.

The project started with an information acquisition stage. Two thousand households in six European countries participated in a quantitative survey run by an external research organisation. In addition to quantitative market research and trend analyses, Whirlpool's internal usability organisation within GCD lead by Gigliana Orlandi observed households and interviewed people. The team, which consisted of a cultural anthropologist, a usability expert and their assistants, observed the clothing care routines performed in people's homes. The information gained through observation was often limited to person–machine interactions. The overall picture of the processes and the needs emerging from them would have remained hidden if the quantitative research had not been combined with GCD's qualitative observations.

Whirlpool Europe's GCD department and three external design groups created the project concepts. One of the key areas of interest – beyond creating the prototypes – was exploring an organisation model that blurred the distinction between in-house design and external design.

The goal of the project was to create research-based designs to function as a platform for discussion and decision-making, not as plans for specific future products. The results of the user survey and information about emerging technologies were distributed at the project launch meeting. The teams became familiar with the information and with each other, and then started developing the design approach. The designers visualised the research findings and processed ideas for alternative washing methods based in particular on the users' experiences. Therefore, the aim of the project launch meeting was not to dictate the design assignment but to allow it to be developed jointly.

Next the teams developed, refined and presented their suggestions to each other using drafts, animations, storyboards and mock-ups. Together they selected the proposals for further development and then continued working to transform them into prototypes. Essentially all the information was shared openly amongst the teams. Jones did not want any unnecessary competitiveness among the teams, considering it more important that they viewed themselves as the Project F team. This collaborative process resulted in five designs that were taken to the prototype stage. Some of them combined existing technologies and some focused further into the future by anticipating new technical solutions.

The prototypes aroused interest amongst the media. The models were showcased widely, for example at the Milan Furniture Fair. In order to obtain as much feedback as possible about a design, it must be presented in a tangible and credible way. It was important at the outset that the team thought through – as part of the design brief – the communications and information design aspects of the project as much as the prototypes themselves. People respond to what they can see. According to Jones, a better perception will produce a better response. A prototype, that includes a description of how the product is used increases its impact. It is also easier for the media to present tangible ideas to the public.

Project F, like any other initiative at Whirlpool, started with a defined timeline, set of deliverables and budget. While these requirements put the project into a "sandbox", Jones emphasises it was important that all concerned understood that the sandbox was a big one – the team needed to deliver something tangible at the end. The design work for Project F was carried out over the course of about 3 months during the summer of 2002. Several months of research work had preceded the design process, which was followed by the construction, photographing and publishing of the models. At Whirlpool it takes between 9 and 12 months to implement a design project of this sort and another 12 months to communicate it.

A good example of Project F's results is the BioLogic (Figure 1.6), which is the prototype that received the most publicity in the project. The BioLogic was designed by Whirlpool's own in-house team, and is based on an environmentally friendly, slow-wash principle. The machine does not have a washing drum. Instead, the water circulates between three areas of the machine. It flows from the storage tank into removable pods where it cleans and rinses the textiles. The water then moves from these pods into a purification tank where hydroponic plants clean the grey water by absorbing the nutrients they need. The water then flows back to the storage tank. The energy required for the water circulation comes from fuel cells, and the heat and water they produce is recovered.

According to Jones, carrying out Project F or any design exploration exercise provides the following benefits. First, it influences on the internal motivation for the design team by giving them a chance to dream, to explore, to stretch their imagination and to challenge their skills. Second, it provides an opportunity for the organisation to experiment with new talents and new ways of working. Not only did Project F promote a dialogue about the role of design, it also brought the company's designers, engineers and decision-makers as well as the in-house and external design consultants closer together. Third, these projects bring external visibility to the company and contribute to brand building by reinforcing the company's reputation as a leader in design and innovation. Project F was a very publicity-oriented project, where the general level of interest was an important criterion in the selection of the designs to be completed. Projects like this convey a company's interest and innovativeness, which boost market, customer and investor interest and help in recruiting.

Since Project F Whirlpool has executed similar design exploration projects. The company sets up a global team of internal and external designers, equips them with the latest consumer research available, gives them the tools and the budget and direction to execute, and then lets them get on with it.

1.5   Three kinds of concepts

A common and characteristic feature of the product design-related functions of KMY Technology, Ed-design and Whirlpool described above seems to be that they produce descriptions of products that are not intended for production. Ed-design designed future user interfaces in order to acquire knowledge for forthcoming assignments and to motivate personnel. Whirlpool, in addition to learning, pursued media visibility for the company and visibility for the design team within the corporation. KMY Technology communicated new possibilities to customers to support investment decision processes. The expertise and the cooperation of design and product development, together with their different stakeholders, were essential for

FIGURE 1.6.

Whirlpool's BioLogic washing machine concept, which is based on an eco-friendly, slow-wash principle

photo: Whirlpool

all these objectives to be met, and the final results presented in each case were product descriptions. None of these, however, has been taken beyond the prototype stage or sold to the public. Instead, product design has been used as an instrument to achieve other kinds of business goals, supporting employee training, corporate communications, brand building, technology development and marketing, to name just a few.

The word "product" in product development literature usually refers to an object of exchange brought to the marketplace. When product design-related tasks are carried out without the objective of providing documentation for production that will eventually lead to a market launch, there should be a conceptual differentiation from the core meaning of product design. We call these activities *concept design, concepting* or *product concepting*. This also gives us a loose definition of a concept and concept design, by which we mean designs and design processes that are not meant to be implemented as commercial products, at least not without major additional design work.

Concept design means different things in different lines of business, and different groups attach different meanings to it. In marketing and network communications, for instance, the concept is interpreted as the definition, coordination and management of campaign planning in general, which differs from our interpretation. Designers often refer to the simple sketches they create to allow them to study the appearance and structural alternatives of a product as concepts. The meaning of concept design is further confused by the fact that concepts are frequently categorised into product design subareas, for example mechanical design, industrial design or user-interface concepts. The emphasis in each of these areas may be on different characteristics and design processes. In this book, we are looking at product concepting, unless stated otherwise, as a function combining several product design perspectives, which means that we are focusing on integrated product development (see for example [3]). We will not widen our approach much beyond the product concept; for example business concepts are not covered in our discussions.

Concept design utilises the practices of product design, including creativity, a user-centred design approach, wide-ranging investigations and design using images and tangible models (the methods and processes are discussed in more detail in later chapters), which provide a more extensive coverage of the company's different functions than product design. It seems that concepting can play several different roles in providing support for a company's business, depending on the business and organisational environment where the concepts are being created and used. On the one hand the projects we have studied and which we describe as having a concepting approach overlap with product design, whilst on the other hand some concepts seem to be closer to science fiction than to engineering. Using our examples as a basis and depending on the work involved in implementing the concept, we can roughly divide concept design activities into three categories as described below. These categories are in part overlapping, but help in explaining the different business roles of concept design.

In conjunction with product development, concepting refers to the fundamental outlining of a product carried out during the first phases of product creation or at least before the design specification is frozen. This interpretation of concepting is well defined in product development literature (see for example [4]). The work makes a major contribution to the later phases of product development, and concept approval is a well-integrated part of the product development gate model used by the design organisation. These concepts – let us call them *product development concepts*[3] – support the definition of the product specification, which is needed to set detailed goals for the design of the subsystems of the product and for the following phases of the design process. When the specifications are drawn up, ideally no commitment should yet have been made to implement any particular concept or solution, even though it is highly probable that one or more of them will be chosen for further development. In practice, a few set restrictions typically already exist. The strategic decision on the new product's position in the company's product portfolio has however already been made when product development concepting begins. Some characteristics of product development concepts can be seen for instance in our case studies of Decathlon's Imaginew process (see Chapter 3) and Nokia's mobile phone developments (see Chapter 5).

Emerging concepts are created in association with technical research or the modification of products for radically different markets. With emerging concepts, the opportunities of a new technology or market and growing user needs are unravelled, made understandable and used to support the company's learning and decision-making processes with regard to future

---

3 These terms are taken from a study by Kokkonen et al.[10], although there are some differences in how they are applied.

product generations. The focus of the examination can vary from the development of individual product characteristics to the identification, brainstorming and development of completely new types of product and service concepts. The design of emerging concepts, which includes research and prototyping activities, is often a long-term process in which projects can last months or even years. Within the framework of product development there is insufficient time for this type of learning. However, it may be possible in the short term to exploit some of the intermediate results or spin-offs, such as new product features, new solutions based on existing technology and targeting new user groups, or innovations based on new manufacturing technology and affecting the availability or price of the product. Examples of emerging concepts are KMY Technology's cruise ship concepts and the concepting project for Nokia car phones (see Chapter 3).

While emerging concepts are based on very realistic assumptions about the future development of markets and technology, there are also concepts that go a stage further and include factors that may not be impossible but which have not yet been proven to be feasible. These concepts are sometimes developed to support the company's strategic decision-making by outlining the future beyond the range of product development and research activities. There is no expectation that this kind of vision concept will be implemented, and therefore the technical and commercial requirements are less restrictive than those in the concepting activities described above. Future research, scenario development and technological anticipation can all lie behind vision concepting. A typical feature of vision concepts is that they are showcased publicly at trade fairs and in the media. In this way visions are used to outline and communicate the company's brand and future strategy. Examples of vision concepts are Whirlpool's BioLogic washing machine and Ed-design's future scenarios.

Our main interest in this book is to discuss the practices related to emerging and vision concepts, which are usually created in concept design or research projects. Product development concepts are produced as part of product development projects, which are extensively described in R&D textbooks. However, the roles of concepting that we have described are not completely separate. Concepting carried out during the course of research feeds ideas for new solutions or new projects into the product development process. In some cases, a concepting project can start as a research activity and continue as a product development project. As is shown later, the learning process and the realisation of products are sometimes so tightly intertwined that they cannot be separated. This is why all three types of concepts must be included when discussing concept design.

1.6　The general objectives of concept creation

Product specifications developed without the goal of a market launch may seem somewhat frivolous. The case studies described above give an idea of the potential uses of concepting. However, the questions of why are concepts created and what benefit companies can expect to justify their investment in concepting still remain unanswered. On the basis of the previous examples and those in the following chapters, it is possible to isolate a few objectives. The case studies used to identify these objectives are taken from the high-technology and device manufacturing industry. On the basis of discussions with and responses received from other industries, such as software development, new media, delivery services, the construction component industry and low-technology consumer durables, we are convinced that the objectives are largely generic and apply to a variety of industrial disciplines.

The above-mentioned objectives link design with several business functions and present it as a highly versatile asset that companies can use for several purposes. When given the opportunity to work in other contexts, design, which has been regarded as a support function for product development, can find meaningful business roles that take it a long way from its origins. The repositioning and redefinition of design as illuminated by the concepting approach is perhaps the most important individual result of our studies. If this is implemented more widely, it has the potential to influence how companies organise their design functions and how designers are trained.

It is worth noting that the practices and related objectives described below, even though they perhaps appear unfamiliar from the point of view of conventional product design and engineering, are not hypothetical benefits that could be only achieved under specific imaginary or exceptionally favourable conditions. Neither do we intend to portray them as normative practices. They are simply recognised goals and practices of concept design in companies that we have studied and worked with. They illustrate what these companies have been achieving or aiming at with their integrated concept design projects. It is clear that all the companies in question have highly developed design functions.

1.6.1   Concept design for product development

Product design has for many years included activities whose main aim is not to try to solve the design problem, but instead attempt to define the design challenge and map the alternatives. In design waterfall models, the first or second phase is typically related to unlocking the problem. In well-known product development textbooks, for example [4], this work is called concept design. We have already categorised its results as product development concepts. During product development concepting, the main outlines of the design are defined, and then details are added during the subsequent concurrent product development phase. The aim of concepting is to prepare for concurrent engineering by specifying the fundamental solution to the design problem, which is used as the basis for the decision to go ahead with the detailed design.

1.6.2   Concept design for innovation

The product design environment often leaves little room for new creative ideas. Creative solutions, inventions and the process of refining them into innovations are welcomed, but the uncertainty that often accompanies them tends to favour the traditional approach when deciding which solution to choose in a product development project. Product design and production start-up can require major investment and can lead to significant financial implications in the event of a solution being unsuccessful. Risks can be managed by further developing and testing new solutions, but the tight schedules of product design rarely allow for the examination of radically new proposals. For example, product development projects for complex mobile technology products often only last a few months. Even if it were possible to respecify one subarea of the product design, the links between the product's different structural groups and subsystems are typically so tight in concurrent product development that respecifying one part leads to other required changes. Standard interfaces between subsystems (i.e. product modules or product platforms), allow changes to be made within specified parts of the subsystems, but at the same time they determine the product architecture. The modifications permitted at the top of platform-based product architecture seldom fulfil the criteria of radical innovation.

Sometimes the changes and their domino effects can be very difficult to forecast. When a tyre manufacturer changes the profile of its tyre, how does it affect a car's cornering characteristics? How is mobile Internet usability affected by the different pricing structures of operators? It is easy to understand why some project managers and project steering groups favour the traditional solution. And if the changes are implemented without being properly integrated with the entire product architecture, it is just as easy to understand the frustration of users struggling with the partially incompatible systems they have purchased.

However, the development of products through substantial improvements, with new solutions that challenge the entire essence and technological platform of the product, is a key means of achieving a competitive advantage. The shift from analogue to digital technology has transformed telecommunications. What would the change from wheels to legs do for the tractor and automotive industries? In many business sectors a radical move to a new platform is a prerequisite for market survival. Consequently, it is essential to be able to implement design work that breaks away from the current concrete, technical restrictions and compatibility requirements set by legacy products and production processes, and from the short-term profit targets set by sales departments. There must be an opportunity to ask "what should we design?" In addition, the conditions for generating new ideas through creative design should be created without the immediate exploitation of the ideas being the initial and most important criterion dictating their final evaluation. This need can be met by launching different concepting projects, the key significance of which is to create the prerequisites for innovation.

Research and technical development create the foundations for product opportunities, but do not identify them. In order to find and implement these opportunities, both insight and design are needed. Design can provide the link between the pushing mechanism of technical development and the pulling mechanism of the market, and is essential for transforming inventions into innovations and for linking the often implicit demand with the emerging possibilities. Design can develop what technology allows into concrete and specific proposals, and it can also be used to sell the innovation. The customers and users of new types of products are not the first people who need the technologies to be understandable. In addition the decision-making process about using new platforms, investing in further technical development or entering new markets with a radically new offering requires concrete proposals where opportunities are clearly articulated.

When interesting technical, commercial, user or vision-design ideas have been generated, there are different ways of making use of them. These include, for example, direct product improvements, idea banks, directing technology development, launching strategic cooperation, and patenting.

While improving current products is not the goal of concept design, it can be a desirable by-product of the process. When design is guided by user and customer centricity, it often deals with needs that can be fulfilled using innovative but easily implemented changes to existing products. The concepting results can also form the basis of an idea bank. At the time when the project was completed, alternative concepts may not yet have been linked into the right context, for example as a result of underdevelopment of the technology and markets, but the idea bank may prove to be useful later when the technology and markets evolve. Those working on engineering projects can consult the bank to find solutions to their design challenges. Concepts that are commercially attractive but not technically feasible and yet not too far away in the future may guide technology development investments. An example of this is the development of shipping technology, which initiated the concept of an on-board promenade. Concepts can also give rise to products that would require new kinds of strategic cooperation between different players in order to be realised. This cooperation might, for instance, be a shared responsibility for technological development, a joint specification of industry standards for the entire business sector or the creation of a distribution network for new markets. Concepts can also help in the identification of key technologies and solutions, allowing them to be patented for a company's eventual use – or to prevent competitors from using them.

Time is needed to find ways to exploit concepts, and here concepting for innovation must look further into the future than product design. It must visualise the outlines of future products before it is pushed through the tightly scheduled processes of product design projects. Concept design projects vary considerably in length, but technology-intensive development tends to be particularly time-consuming. And as it is important to be ready early and you may need plenty of time, you cannot delay the start.

1.6.3 Concept design for shared vision

Changes in the business environment require decisions about the product line to be made continuously. These can include adjustments to product specifications, the introduction of new product variations, the selection of the product platform and making use of new technologies and markets by means of completely new product lines. These decisions affect a large group of different players at different levels of product design and corporate management. Decisions must be based on information from market surveys, technical research and development work. In decisions related to product lines, products and their subsystems, the information about markets and new technology – even if it is accurate and credible, which is not always the case – cannot be easily exploited. It deals with premises, while the decisions concern designs and concrete actions.

The conceptualisation of new products makes the alternatives more tangible. It enables questions to be formulated more accurately, gives an understanding of what kind of information is needed and provides an insight into the meaning of existing information. Conclusive decisions about a product line based solely on the background input data with no understanding of the potential tangible products are on shaky ground. As the Whirlpool case described above underlined, examples – that is concepts – are needed of the kinds of tangible new products that technological developments and market challenges can generate.

The greater the amount of interdisciplinary and cross-cultural communication in the decision-making process, and in the global consumer durables and business-to-business marketplace there is a great deal, the more extensive and demanding the discussions needed to achieve shared understanding. Unfortunately, the products of the future are often described – and the corresponding decisions made – based on very obscure expressions. For instance, commonly used expressions such as "advanced" or "easy-to-use" may describe a desirable product in such a way that everybody can agree on them around the meeting room table. They are, however, much too vague and too subject to individual interpretation to be used as reliable foundations for decision-making. What kind of product and user experience will they result in once they have been interpreted by an external design or engineering consultancy? How advanced should the product be before it starts to become bizarre? For whom should it be easy to use? A software developer perhaps? How easy can it become before it starts to be trivial?

These words and several others are used lightly as if there were a shared understanding of their meanings within the organisation, while in reality there is not. A false image of a shared goal is probably much more dangerous than the realisation that it does not exist. Concepting through simulations, pictures, diagrams, stories and mock-ups transforms existing possibilities, wishes, abstractions and words into physical form and creates a shared foundation for understanding.

At best, the concepts of future products are not disconnected guesses, but instead form a broader image of the future. With several alternative proposals organised according to understandable principles, concepting can create a map of the future and add landmarks to it. The map can then be used to position hypothetical products in the foreseeable future, to make use of the future technologies that they incorporate and to offer them to hypothetical customers. The concepts act as landmarks in the future terrain that make navigation easy and intuitive. When the concepts are given memorable names, the map also defines a vocabulary for the written and spoken language. It provides the organisation with new words with well-specified meanings tailored to its communication needs.

On the basis of concepts that have been created, future products can be discussed in just the same way as those that already exist. It is possible to start assessing their characteristics and even to start liking them. In the best-case scenario, they can be used directly as models for products that will be produced. However, even if none of them are implemented, the landmarks in the future allow a more specific and more expedient company approach to be defined. By presenting cruise ship concepts to its customers, KMY Technology created a new language that customers and the shipbuilder were able to use to discuss the future of the cruise-line business and its related products in precise terms.

1.6.4 Concept design for competence

Concept design prepares companies for future solutions by exploring alternatives. The preparation process may in some cases be linked to the company's own expertise, as well as or instead of to future products. In fact in most of the cases we have studied, learning has been, if not the main objective, at least an extremely important secondary objective.

A product design company must maintain its innovation potential. Creative design is a skill that can be learned at both the individual and organisational levels. For example, IDEO, one of world's leading design firms, emphasises that creativity is part of an organisation's expertise, and that it is not coincidental or mysterious[6]. In a similar way, skills can be lost. Routine product design that offers no opportunities to challenge one's own expertise can cause that expertise to fade. As processes to make work more efficient are developed, the ability to step back and to see the world and one's own role in it from new perspectives can be lost. It is precisely this ability to change perspectives that is a critical skill and the sign of a good designer[7]. Designing something that no one has asked for, defining the design challenge and responding to it are ways of breaking the routine. Concept design does not have to succeed by being perfect in the same way as product design. This is why it is possible in concept projects to learn as a result of bold forays and the failures that sometimes result from them. This is also why many exercises resembling concept design are carried out by trainee designers at design colleges. And this is why it is also beneficial for very experienced design experts to occasionally step out of the confines of the box and take a concepting approach, just as the design departments of Ed-design and Whirlpool did.

Permission to fail, which can be applied during an individual's learning process, is a key that can also unlock doors to the development of new forms of cooperation outside the department, company or even industry. Companies, research centres, educational institutions, customer representatives, technology suppliers and representatives of government bodies can study emerging phenomena together, compare views and establish connections within concept design projects where there is relative freedom to act. They can take the time needed to learn about each other's opinions, about the language they use or even to develop a new shared language as described above. In a delivery project, direct business interests can prevent wide-ranging cooperation because of tight schedules, legislation on competition and confidentiality issues. In contrast, concepting is free of these restrictions. Positive experiences of cooperation may, in addition to improving skills, result in heightened self-esteem and team spirit.

In addition to promoting creative skills and cooperation, concepting is an ideal framework for learning about new technologies and business opportunities. It forces teams and individuals to make use of new information instead of just scanning through it. This is an excellent way of gaining a genuine understanding of the new material. When you have to create a new solution based on a new technology, you must have an in-depth understanding of what it means to your organisation, your customers and your business. During the concepting process you will come across misconceptions, make mistakes and identify the dead ends so that you can avoid all these issues when the stakes are higher.

1.6.5   Concept design for expectation management

Once you have seen an impressive product, it is impossible to forget about it. When such a product is launched on the market, it changes consumers' ideas about what they want or think they need. Typically consumers only realise that they need something after they know that it exists. In extreme cases, the effect can be instant and categorical, but more often people become accustomed to new products gradually as they make their way into mainstream society and culture.

Seeing a product being used by other people and portrayed in the printed media and advertising contributes to consumers' perceptions about how well accepted or necessary the product is[8]. However, the process of bringing new products into the mainstream can begin even before the product is on the market. Companies can control the ideas and expectations of the general public through communication. Sometimes it may be expedient to prepare people for new products by showcasing them in public before they are actually ready. Even a radical product that is presented in advance as a concept may be exactly what consumers have come to wish for when it is launched a few years later. The BioLogic slowwashing machine would be too bizarre to purchase if it suddenly appeared in domestic appliance stores. Whirlpool is creating a soft landing for consumers in the radically new technology by presenting the BioLogic concept long before it tries to launch it – if it ever does. Preparing consumers to accept new designs is an established practice in the automotive industry, as is described in Chapter 5.

The presentation of interesting concepts is often linked to more generic business goals than those closely related to the specific designs. Such concepts aim at improving brand recognition and reputation. Communicating interesting product concepts can help to build market expectations and also contribute to the company's or the entire industry's image as a promising career alternative or a lucrative investment target. As described above, KMY Technology presented new ship concepts in order to explain the dynamic developments taking place in the shipbuilding industry. In this context concepts can be very cost-effective, since the media have an endless interest in new products and features. Pictures and articles will be published without the company having to use its marketing budget for advertising, provided that the presentation material is eye-catching and convincing.

The realisation of radically different products often requires the participation of several companies. This is necessary when the companies are operating in markets where the product systems consist of compatible technologies from several suppliers. New physical products need to be supplemented by software, services, biotechnology, media content, etc. Development work is often shared between different stakeholders, such as companies focusing on their own core competencies, but also universities and research facilities. The future business related to the new products must be made visible, so that all the activities that will enable the realisation of the concept will begin to emerge around it. Therefore it is important to communicate the opportunities and to make them attractive in order to entice the critical mass of participants. Industry-leading companies target these

design for **product development**
  • specification for the following design phases
  • decision to go ahead with implementation

Concept design for **innovation**
  • spin-offs for immediate improvements
  • idea bank for future use
  • concept directions for technology development investments
  • alliances with key partners
  • patenting

Concept design for **shared vision**
  • specific shared meanings
  • vocabulary for communication

Concept design for **competence**
  • improving creative problem solving
  • improving cross-disciplinary -cooperation
  • learning about technology and market opportunities
  • improving team spirit

Concept design for **expectation management**
  • improving brand image
  • influencing consumers' acceptance level
  • influencing stakeholders' interest

FIGURE 1.7.
The objectives of concept design

processes in the direction they want to take by revealing as concepts the products that they are planning to bring to the marketplace in the future. The smaller players adjust their activities to align with the bigger ones, ensuring that the future indeed becomes just what is needed for the leading players to succeed.

Concepts can change people's attitudes and expectations, which can lead to changes in behaviour. Concepting is a tool that can influence attitudes and consequently influence the future world. Figure 1.7 summarises the objectives of concept design.

1.7   The properties of a concept

The objectives of concepting, as listed above, resonate deeply with the principles of product development and business operations. The interface between concepting and business objectives is different to and broader than that of product design. So what kind of output is the concept that promises so much from design alone without manufacturing? Different objectives emphasise different characteristics in concepting and in a concept, but it seems that a concept that can achieve several of these objectives should be a description of a product (or service) that is:

- anticipatory
- well-founded
- focused
- understandable

It would be tempting to give a generalised and simplified definition of concepting as a design activity that looks further into the future than product design. However, based on our examples, it is not easy to specify a time span that would be typical of anticipatory concepting and which could then be compared to the anticipatory perspective of product design. In addition, when product development directly serves production, this is often based on long-term presumptions about future markets and user needs, even though the technology may be based on existing information. This is essential at least when the products have a long life span, as is the case with many investment products (which can be used for decades). When operating in a very uncertain and dynamic business and technology environment such as information and communication technology (ICT) has been, the concepting type of activity is associated with a quick market launch of products, which means that the concepting function has a very short time span. Here concepting aims not to anticipate the long-term future, but to put out feelers at that moment to sense the rapidly approaching and yet uncertain future. In the more mature and established industry sectors, such as the automotive industry, product development cycles are longer, and concepting even relatively realistic visions can take several years from concept creation to realisation.

Therefore, giving specific time frames does not help us to understand the future orientation of concept design. Including "anticipatory" as one of the attributes of a concept simply means that there has to be enough time to utilise the concept for making and implementing the decisions – whatever they may be in each case. Concepting should be able to contribute in a proactive manner by helping the company to take the initiative. If the company simply reacts to what others do, no matter how long the time span is, concepting has not been carried out properly. It must be possible to use an anticipatory description to create product specifications for the product design platform so as, to influence expectations of a product before its market launch, to steer technology development or to learn for future assignments. At its furthest extent, concepting can be used to anticipate the future of the entire industrial sector in an attempt to make it as favourable as possible for the players. The time span of the anticipation (e.g. months or years) depends on the sector, the evolution of technology and markets and the length of product development cycles. It is easier to anticipate further into the future in a sector with mature technologies and markets than it is in a dynamic and uncertain environment. Comparing concepting in the automotive industry (Chapter 5) and in very uncertain business environments (Chapter 6) sheds light on the different concepting and product development cycles.

A well-founded description of a concept supports and promotes understanding of the solutions presented. In order for a concept to function as a tool for decision-making and anticipating the future, it must be associated with the phenomena that are presumed to be significant in the future. A concept must take a stand on the presumed needs and preferences of users and the opportunities afforded by the technology. Often supporting a concept with convincing arguments requires a preliminary evaluation of it to be made. The main challenges of further developing the concept need to be identified. The fact that a concept is presumed to be well-founded does not mean that it must be proven to be right. That would not be possible, as all product and business decisions include unknown variables and are inherently risky. When presenting a concept, it is essential to highlight the new possibilities. If they are sufficiently attractive, the studies can be intensified and the feasibility of the concept determined during the process of further refining it.

The objective of concepting is not a product description that comprehensively defines the product. When creating a concept, it is sufficient to focus on the characteristics that are fundamental to the product. This is sufficient for decision-making purposes and can save a considerable amount of time. The fundamental solutions are those that distinguish the concept from existing products or other concepts. The characteristics may be related to the main functions or to the benefits they bring to the user, such as the dimensions of the usage experience, the appearance, style, ergonomics, interaction, key technologies and their maturity, the product's size, or the intended user segment. The description must include the characteristics that are critical to the objectives of the concepting project, but at the same time be as concise as possible so as to allow for fluent communication and easy understanding. This also enables the concept to be updated on the basis of feedback received. On the other hand, testing the concept and communicating it may require finished models and prototypes.

In most applications, the concept functions as a communication tool and enabler. The proficiencies of those interpreting the concept and those making decisions based on the concept may differ quite dramatically from the proficiency of those who created the concept. When using concepts to build and probe customers' and other stakeholders' expectations, it is essential to present them in public. Therefore the concept must communicate its message in an easy-to-understand format. The forms of presentation that are often used include usage scenarios set in a story format, 3D models, simulations, metaphors and comparisons describing the nature of the product and the goals of the design[9]. In addition the name, as discussed above, is an important part of the concept presentation as it acts as the handle for meanings that the concept refers to.

1.8 References

1   **Keinonen, T., Andersson, J., Bergman, J.-P, Piira, S. & Sääskilahti, M.** (2004): Mtä tuote-
    konseptointi on? [What is Product Concepting?] in Keinonen, T. & Jääskö, V. eds. Tuotekonsep-
    tointi. Teknologiainfo Teknova, Helsinki

2   **Cooper, R.G. and Kleinschmidt, E.J.** (2001): Stage-Gate process for new product success.
    Innovation Management

3   **Cagan, J. and Vogel, C.M.** (2002): Creating breakthrough products – Innovation from product
    planning to program approval. Prentice Hall

4   **Ulrich, K. T., Eppinger, S. D.** (2000): Product design and development. 2nd international
    edition. Irwin-McGraw-Hill. Boston

5   **Rittel, H. W. J., Webber, M. M.** (1984): Planning problems are wicked problems, in
    Developments in design methodology, Ed. Cross, N. John Wiley & Sons, Chichester, pp. 135-144

6   **Kelley, T.** (2001): The Art of Innovation: Lessons in Creativity from IDEO, America's Leading
    Design Firm. Doubleday, New York

7   **Lawson, B.** (1990): How designers think – the design process demystified. Second edition.
    Butterworth Architecture, London

8   **Pantzar, M.** (2000): Tulevaisuuden koti: Arjen tarpeita keksimässä. [The Invention of Needs
    for the Future Home] Otava, Keuruu. (In Finnish)

9   **Lindholm, C., Keinonen, T.** (2003): Managing the Design of User Interfaces, in Mobile
    Usability: How Nokia changed the face of the mobile phone, Ed. Lindholm, C. Keinonen, T.,
    Kiljander, H. McGraw-Hill, New York

10 **Kokkonen, V., Kuuva, M., Leppimäki, S., Lähteinen, V., Meristö, T. Piira, S. and
    Sääskilahti, M.** (2005): Visioiva tuotekonseptointi – Työkalu tutkimus- ja kehitustoiminnan
    ohjaamiseen.[Vision conception – Approach for Guiding Research and Development]
    Teknologiainfo Teknova, Helsinki. (In Finnish)

2

The Concept Design Team

Turkka Keinonen

2

The Concept Design Team

Turkka Keinonen

We consider a team to be the unit that creates concepts. Individuals, whether they are inventors, designers, design engineers or visionaries with remarkable skills from a variety of professional backgrounds, take on the role of team members in order to develop concepts. An organisation, however creative it may be, does not create concepts, but provides the environment for concept development and facilitates the process.

Why do we hold this view? The creativity and skills of talented and well-trained individuals are essential resources. The creative potential of organisations as whole, including all their members, is well-recognised but less widely used[1]. The first and the most straightforward answer is that in all the cases we studied for this book, concepts were generated by teams. The second answer is that contemporary literature about products and innovation shares the same view. Professors Jonathan Cagan and Craig M. Vogel[2] at Carnegie Mellon University highlight the importance of integrated product design where engineers, marketing experts and designers work in close cooperation to recognise product opportunities and transform them into products. The team has even been called the backbone of innovation[3].

The size and set-up of the teams in our examples and in the relevant literature naturally vary considerably. In some cases a very small team has been responsible for carrying out the vast majority of the activities of a concept project, while in others the process has involved almost the whole product creation organisation of the company or division or has even spread beyond that by involving experts from several organisations. Nevertheless, the tasks were carried out by teams of people who contributed their complementary knowledge, expertise and skills which together enabled the project to be completed. According to Katzenbach and Smith[4] a team is "a small number of people with complementary skills who are committed to a common purpose, performance goals and approach for which they hold themselves mutually accountable". This definition is also particularly appropriate for concept design teams.

There is a huge amount of literature about teams and team management. This chapter aims to give an overview of the team as a concept-generating unit by discussing the composition of a team, the roles of individuals in a team and the leadership challenges of a team. It will become clear that concept design teams and their requirements for their productive work are not in essence different from other teams that are responsible for creative problem-solving. However, in our attempt to describe concurrent concept design activities, the point of view of the main player – the team – must not be forgotten and therefore it is important to look closely at the role of the team.

2.1   Team members

When we speak about teams we are actually speaking about packages of complementary skills. The individuals may be talented, but if the skills within the team do not complement one another and correspond to the challenge faced by the team, their talent will be of limited effect. Even less important than individual excellence is the status or hierarchy within a team. If the team needs formal authority to be able to proceed or if the team members need authority to justify their opinions, very little can be expected from the project. Therefore, an efficient concept design team should consist of individuals with varied and complementary skills, who work together towards a shared goal without carrying too much extraneous baggage. They do not necessarily need to work for the same organisation, sit next to each other or already know the other team members, but this may be advantageous.

As a result the correct composition of team is a critical to successful concept design. The team must have sufficient expertise in the fundamental areas of the product that is being concepted, the right personalities and the skills to acquire the information needed to support the team's tasks, to creatively interpret the product and to realise the concepts. The key expertise must be available within the team, while more peripheral, supplementary contributions can be brought in as needed.

Even though different product categories and businesses naturally call for different competencies, there seem to be certain profiles that are required in most concept design teams. The available literature contains two main approaches to forming teams: one addresses the necessary professional expertise of the team members and the other looks at the individuals' problem-solving styles and social behaviour. The professional expertise approach usually requires engineering, marketing and design disciplines to be present (e.g. [2]) for generic new product design projects. For more focused projects the composition may be different. In user interface and interaction design, engineers and designers work together with psychologists and social scientists (e.g. [5]).

There are numerous models for psychological screening criteria to ensure a good combination of problem-solving styles in a product development team[6]. The models categorise individuals on the basis of their psychological characteristics to allow managers to put together a team with the right balance in several dimensions. Extrovert, action-oriented team members should be balanced with those with a more introverted approach who prefer to complete tasks on their own. Analytical thinkers, who need to understand the reasons behind an action, should not dominate or be dominated by intuitive team members who trust their insight, whilst rushing after attractive but hopelessly complex options. Originators, who bring in new ideas and introduce radical changes, should be complemented by conceivers, who prefer more gradual improvements and want to bring projects to a successful conclusion. The different styles of the team members should complement each other in the same way as their professional backgrounds. Productive tension between the team members with different styles is often seen as a necessary component for innovation.

A third approach to identifying concept team members, in addition to their educational backgrounds or business functions and personalities, is their responsibility within the project. The team members can be categorised by the role they play in the project. A role is something that an individual can adopt for a specific purpose or within a specific context, assuming that he or she has the necessary qualifications. In different types of project the same individual can take on different roles. By choosing an alternative way of describing the team members, we are not implying that the other methods do not make excellent management tools. It is clear that specific educational backgrounds and personality traits are needed by people who take on certain roles. The roles in a concept design team are:

- user expert
- domain expert
- design expert
- communications expert
- feasibility specialist
- team leader

2.1.1 User research expert

The user-research[1] expert is responsible for ensuring that the team has an appropriate and adequate understanding of the user and the context in which the concept the team is designing will be used. In user-centred concept creation, in which by definition there is a continuous dialogue between the developer and user communities, this is a key role. (User orientation is covered in more detail in Chapter 4.) The concept development team cannot trust in and build on only an organisation's existing knowledge or on secondary information sources, such as research articles or commercially available trend forecasts. Instead, the team must have direct first-hand contact with

---

1   The terms used to describe the person for whom the future concept is intended varies depending on the point of view of the author and the relevant research tradition. "User", "customer" and "human" are the most commonly used expressions. "User" emphasises the interpretation of the person as someone who is in operational and goal-oriented interaction with the product, such as "uses" a mobile phone to call someone. "Customer" underlines the decision-making behaviour when purchasing a new product and the more long-term satisfaction with the product, where various issues that are not directly related to the attributes of the product may have an influence. A "customer" chooses the mobile phone but can become dissatisfied with it when her friends replace their phones with more impressive models. "Human" highlights a holistic perspective on the person for whom new products are designed. The problem with "human" is that it fails to explain the difference between the roles of different players in the concept design team (e.g. designers are also humans). Here we have chosen the term "user" because it defines the role of the person for whom the team is designing the concept with reference to the concept. User orientation in concept design is covered in more detail in Chapter 4.

the user community in order to learn in a focused way about users' needs and to create a channel for immediate responses. The user expert is not necessarily familiar with the specific users, practices, contexts or cultures of use in the current project, but he or she understands the methods and approaches used to gather the information. To make the project genuinely user-centred, the user expert must involve the whole team in the dialogue. Consequently, this expert must make an effort to help the team to create a communication link with the users and to interpret the data from several angles.

User experts can be psychologists, sociologists or anthropologists who are familiar with the product development environment. However, they are often also design engineers, product designers or interaction designers who specialise in user research or are familiar with usability design techniques. Analytical skills, excellent organisational capabilities and a certain amount of empathy are needed. For the users, the user expert is a newcomer who is learning about their world, whereas for the development organisation he or she is a messenger who is a living reminder of what is happening outside the team[7].

2.1.2   Domain expert

The domain expert is the person who is familiar with the activities, the products and the markets for existing products in the domain for which the concepts are being designed. Companies often have domain experts readily available in their marketing departments working as product or cat-egory managers or in customer service either in direct contact with the field themselves or at least with contacts within their customers' organisations. However, domain experts need to be found elsewhere when the concept design addresses completely new areas of business and human activity. Nurses and doctors can be domain experts for the design of medical equip-ment, while for sports equipment the experts may be professional athletes and serious amateur sportsmen. The domain expert works with the design team and shares the goals of the team. This is why his or her role is differ-ent from that of a user, whose needs are studied by applying user research techniques and who is not a member of the team. In integrated co-design practices the role of the domain expert and the user can merge.

Domain experts have a considerable amount of knowledge of the activities and culture within which the design will be positioned. However, trusting a single person to represent the user communities is always a questionable approach. This is why the essential role of the domain experts is to work together with the user expert and act as a guide to the new cultures and human activities. They indicate where to find the lead users and other essential information sources. They know how to frame the user research objectives so that they make sense from the point of view of the specific practices of the activity. The sports-monitor manufacturer Suunto Oy has chosen the strategy of "hobbyism"[8], which means that the company recruits new employees based on their expertise and encourages existing employees to take the role of the domain expert by becoming deeply involved in certain sports that the company considers relevant. Consequently, the company employs top-level international and national athletes as designers and product managers for their diving, sailing and hiking monitors, which allows the company immediate access to the culture of the sport.

2.1.3   Design expert

The design expert is a professional whose core expertise lies in generating solutions, which may be formulated as product structures, product appearance or interaction. The professional background of the design expert is typically in design engineering, industrial design or interaction design.

One of the most difficult and poorly understood phases of new product development (NPD) is the step from recognised requirements to proposed solutions – the leap from analysis to synthesis. This is where designers' visual presentation skills, enthusiasm for new approaches and ideas and ability to think in terms of concrete solutions become important.

The design expert contributes by generating the solutions. He or she gives a concrete form, order or structure to something that has been an abstraction or, strangely enough, that has not existed in any form before the act of visualisation has defined it. This process of concretisation causes it to come into existence. In order for this to happen it is not sufficient to simply provide the premises. The act of presenting the solution also requires the designer to introduce something extra from his or her own insight and experience. Design experts probably do not have more insight or expertise than anybody else, but they are trained to trust their intuition and act on the basis of the knowledge that they have. This is why they are typically more confident about presenting solutions for which there are only partial justifications. However, designers do not have a monopoly on proposing solutions; all the other team members obviously share the responsibility for bringing up new ideas. Therefore perhaps the most challenging task for an effective design expert is not only to be able to design, but also to enable the other team members to design, to make the whole team design and to turn a research or engineering project into a productive design project. This can be done by producing presentations that everyone can comment on, experiment with and improve. A good design expert helps the whole team to solve problems.

2.1.4   Communications expert

The role of the communications expert is closely linked to that of the designer, but is not identical to it. Whilst the design expert creates the rough mock-ups and sketches for internal problem-solving within the project, the communications expert creates deliverables that communicate the features of the concept to outsiders. He or she prepares convincing, attractive project deliverables, including storyboards, scenarios, 3D models and renderings, photographs of the models, interactive simulations, presentation slides, web sites and trade fair stands. Several of the objectives associated with concepting can only be achieved after the concepts become well known, so a publicity campaign is often essential. The skills required may include a wide range of traditional and computer-aided visualisation and model-making skills. The task of creating the deliverables is often more than one person can handle. Fortunately it is easier to supplement the core-team's expertise when it comes to planning and implementing the communications than it is in other areas.

2.1.5   Feasibility specialist

The feasibility specialist ensures that the concept design team is aware of the technical restrictions and emerging opportunities. When the concept generation process is technology-driven, the technology expert, or in fact a technology subteam made up of several experts, takes the key role in driving the process. In more user-centred approaches the feasibility specialist provides knowledge about current and forthcoming implementation options and carries out an initial feasibility study on the options that have been created.

2.1.6   Team leader

The team's commitment to its goals and its desire to succeed depends to a great extent on the team's internal dynamics. It is possible to produce favourable conditions, create an inspirational mood and generate motivation by leading the process. The leader of the team plays a key role in this respect. The right kind of guidance and the right conditions can also reduce friction, such as unnecessary disputes about irrelevant issues based on personal opinions and egos – this sort of dispute can easily arise in a self-directed team. The leader's other main responsibility is to network the project with other relevant projects, communicate with the stakeholders whose commitment and decisions are important for ensuring the continuation of the project and at the end to continue processing the results. The third responsibility of the leader is to act as an integrator and ensure that all the important issues have been taken into consideration. Therefore, the leader must have an excellent understanding of and a good grip on the project as a whole. The educational background of a concept design team leader depends on the subject of the project. Professional managers are probably the best choice for complex, large-scale projects. Engineers are more suitable for technology-driven projects and designers for projects where the user interaction is the main challenge[9]. For visionary and risky projects, the passion of the project leader is an essential resource; the person most likely to be passionate about the project is its originator[6].

2.1.7   Setting up a concepting team

As described above, there are quite a few skills and roles that are essential to a concepting team, and several of them are broad enough to require a team of their own – not just an individual expert. However, because concept design aims to describe a future product on a rather generic level, very detailed knowledge is not always needed. It is typical, especially in smaller teams, for one person to take on several roles. This is why generalists, multi-field experts and senior experts are good resources for concept design, even though their skills would not be sufficient for the final implementation. For instance, the same person can – and often does – assume the roles of design and communications expert. The domain expert can also network and facilitate team activities by taking the role of team leader.

Figure 2.1 provides a description of the teams that carried out the user-centred concept creation projects covered in detail in Chapter 4. The

Phillip Butt, Suunto

We are motivated by how we can deliver the "want-me factor". This is question we hope all our products will ask the consumer, and it manifests itself in pride of ownership and a sense of satisfaction. Ultimately we want people to want our stuff.

Concept creation is executed in both long- and short-term initiatives with a definite emphasis towards the latter. The "official concept creation", meaning those projects that have research status, tend to be technology-centric and the domain of a scientist. They would greatly benefit from the input of a multidisciplinary team that could contribute depth to the concept, possibly leading to surprising and unexpected manifestations. The design department contributes through user-centred concept creation studies, aesthetic and mechanical design, user-experience design, and more generally by knowing the sports, the people and the business.

The concept creation initiatives I have personally been involved with include outlining a new business opportunity augmenting existing technological competencies, exploring the potential of a newly entered market that had questionable business fundamentals, an online survey that identified mass customisation preferences with our existing market and a study of a new consumption paradigm directly challenging the way we currently segment our offerings.

The number of times we have been misunderstood over what is conceptual and what is production-oriented is simply stunning. Without clear differentiation about the design intent there can be quite a few misunderstandings – these are avoidable.

Younghee Jung, Nokia

Creating concepts for human communication often requires us designers to keep our eyes on both the mundane aspects of life as well as the solemn academic theories. I always joke that my sources of inspiration range from Cosmopolitan to The Economist and Bonnie Nardi's publications.

My responsibilities include identifying what to research on and design for, coordinating the direction and execution of user research by other experts, participating closely in the execution of user research and identifying constraints and trends in business and technology; and, in general, making sense of stuff.

Working in an environment full of technogeeks, we are often approached by people asking us to generate concepts while sitting in a 1-day workshop. A design concept is not born within the space of 1 day. But it is difficult to prove that ideas that are designed are better.

We've created a concept that people can use while in face-to-face discussions. We created a set of user scenarios and ran focus groups. When they were presented with the scenarios, the focus group participants started to say "This is an insult to our social skills". The facilitator did not understand the concept very well and because she was so embarrassed by the impassioned discussion, she did not want to continue. We had to look for another way.

Many people will need some convincing to understand the value – and the concept – of concept design. Sometimes it is not about innovation and creativity, but more about creating a sensible design that can be implemented. Sometimes it is all about finding a ground-breaking idea.

Jane Fulton Suri, IDEO

My role is primarily to tell stories and encourage others to tell stories too. These are stories about real people's behaviour and experiences now, and stories about a future where both the client company and its customers feel more effectively supported. From stories we discover patterns in behaviour, motivations, pleasures, frustrations and opportunities for technological developments to provide a better experience in the future. This is a way of helping teams uncover insights about humanly beneficial opportunities that lie ahead, and it often involves interacting directly and in context with a range of people who represent the edges and extremes in a particular domain.

What motivates me is simply the idea that design is an exciting creative activity that is all about exploring the abilities of technology and human imagination to make life better. For me it is important that conceptual design is not just blue-sky thinking, but is rooted in the overlapping reality of human, business and technical possibilities.

One of the joys of my work is the privilege of learning about other people's worlds, especially the sense they make of technology systems that do not work perfectly for them, by designing their own surprising and delightful ad hoc solutions: a list of important phone numbers written with a permanent marker inside the cover of a flip-format mobile phone or discovering a magazine picture of a favourite television star that a girl propped up in front of her video recorder to remind her to record the show. She would never have learned to set the system to record automatically. These workaround solutions are amusing, memorable and very often convey important themes for conceptual design.

←↑ FIGURE 2.1.
Concept design experts

number of participants, the scope and the scale of the projects vary considerably, but the same roles can be identified.

In Nokia's steering-wheel project (see Chapter 4, and [10]) for in-car product concepting, there were domain experts who came from the company's two business units responsible for factory-installed car products and aftermarket accessories. Their role was to ensure that the requirements and practices of car manufacturing, supply and the accessory business were recognised. They also made sure that the results of the concept creation process were communicated within the business units to the people who needed the information. These people were relatively new to the organisation and hence were supported by more senior mentors.

The role of user experts was played by a mixed team of sociologists, psychologists and usability engineers from the company's research laboratories. They planned the user research and usability evaluations together with the domain experts. The domain experts, for instance, specified the geographical market areas where the studies were carried out, and the user experts chose the methods and organised the field observations. The observations were interpreted in teams involving all the key roles.

The technological expertise for the project was provided mainly by product development engineers working for the business units. The user scenarios, physical designs and interaction were handled by senior industrial and interaction designers. These people also took part in the preparation of communications material, but the majority of this work was carried out by more junior designers, software specialists, graphic designers and model makers.

The project was led jointly by one of the domain specialists and the head designer. A total of around 20 people took part in the project, but the core team who worked throughout the 12-month project consisted of five people providing domain and design expertise. The involvement of the other people was necessary to provide the required expertise, to allow the project to be squeezed into the specified period of time, to improve the cooperation between the different units in the company and to share knowledge, because one of the units had recently been merged with the Nokia organisation.

In the four-wheel kick-bike project (see Chapter 4), the product development manager in the customer organisation was responsible for providing the domain expertise, the technological expertise and for leading the project. The consulting industrial designers simultaneously played the roles of user, design and communications experts. The aim was to create a picture of the factors critical to the success of the new type of product. Since the scope of the project included concept creation rather than a detailed product description, a senior industrial designer was also able to produce the structural solutions under guidance from the development manager.

2.2   Helping the team to work together

As we emphasised during the introduction of the team roles, it is essential for the experts to initially have good personal skills, but these skills must be developed so that they can be used within the team and eventually beyond it. This is the only way in which the interaction can be seamless and fluent, and by which a collection of experts can be transformed into a team capable of integrated development. For professionals – and designers in particular – whose expertise has traditionally been very tacit, this can be a major challenge. Below we will discuss three factors that may have an impact on the team's cooperation, namely the working methods, physical premises, and the size of the team.

2.2.1   Methods for working together

Several interactive methods for creative teamwork already exist, and more are being developed in design research institutions. For the present purposes it is sufficient to mention that appropriate team-based approaches are available for:

- interpreting qualitative data
- building scenarios
- ideation
- evaluation

Some of these are introduced in Chapters 3 and 4.

A good teamwork approach enables the team members to contribute their personal insights and expertise no matter how quiet and reticent they are or, for example, how unconfident they are in using the language (e.g. the case of multinational teams). The methods must also enable the team to build on the individuals' contributions so that one idea or interpretation can be used as a stepping stone to another more advanced one. This is why team methods are typically combinations of individual and shared phases. Well-known examples of team methods that alternate between individual and team phases include heuristic evaluation[11] and the affinity diagram[12].

The approaches should also allow the teams to create a good balance between joint sessions and individual work outside the team meetings. Even though the advantages of team processes have been emphasised, teamwork multiplies the working time of an individual by the number of team members. One person-month of working hours can easily be spent in a couple of team workshops. Therefore, those parts of the design process that do not benefit from the team contribution need to recognised and carried by individuals.

Within the industrial design community there is some mistrust of formal approaches that do not exactly match the designers' requirements. However, in the same team there may be individuals who can take comfort from well-defined approaches during the stressful concept creation process when the results are on the borderline of being achievable. Therefore, the methods should be clearly defined to give the guidance needed, but at the same time they must be transparent in order to ensure that using the method does not become the primary focus of the work. The method should allow the team to follow the approach, but at the same time let the team and its individual members focus on the content of the project. The method must support the process without focusing too much attention on itself.

The methods should provide support for the creation of comprehensive and easily understandable documentation. Documentation plays an important role in the team by concretising the issues and providing a shared ground for reference, annotations, discussion and learning. After meetings and workshops, the documentation helps to remind the team members about the results of the meeting and the paths that they followed to achieve them. However, working documents are rarely of sufficient quality to be used as presentation material or sufficiently self-explanatory to be shown to people outside the team, but in the best case it should be possible to minimally edit them to produce formal documents. A research team at the University of Art and Design Helsinki, UIAH has experimented with video recording all the sessions of a user-centred concept design project. The video has turned out to be a useful tool for briefing new members who join the project about the earlier phases and the decisions that have led to the current project phase.

2.2.2   Physical premises

Concept design teams are expected to propose solutions that go beyond day-to-day engineering problem-solving. To achieve this demanding goal, teams need design premises that are appropriate for concept design. For example, Decathlon left the company premises and travelled to the Alps (see Chapter 3). Putting a physical distance between the team members and their desks, cubicles, hard drives, appointments and urgent e-mails makes it easier for them to leave disruptions behind and focus on creative work. The distance can also help team members to forget temporarily the company's segmentation models, style guides, technology platforms and all the other handy and effective methods for reusing current solutions. Practices that only permit the use of streamlined engineering processes for the on-schedule implementation of products can jeopardise the use of the imagination and the unconscious "tortoise mind"[13]. In this mode of thinking, which is typical in concept creation, it is crucial to allow time for ideas to incubate and crystallise.

Unfortunately travelling to distant locations and dedicating longer periods of time to a single concept design project is often impossible; in fact the opposite seems to be typically the case. Product development resources are often tied to several projects running in parallel and some of these projects are in firefighting mode, which means that concepting simply has to give way. Consequently, concepting is fragmented over time, being carried out when the time allows. When time cannot be used to package the project and keep its activities coherent, other means, such as physical space, should used for this purpose. Therefore, it is advantageous for the project to have an area where the project documents – mood boards, storyboards, sketches, flow diagrams, affinity diagrams, etc. – can stay hanging on the walls to remind team members about the previous steps taken and the solutions proposed by the project. These objects will also help team members to recall the working mood and spirit of the project when they come into the area. Entering a project area can help the team members to remember the unfinished problems and tune themselves into the project more easily than just finding the project folders in the group work application (which probably will also have to be done).

In the case of Ed-design, the managing director's office was used as the project area and the trend forecasts were put up on the walls. It may be that what was lost in terms of an informal and relaxed working atmosphere was regained by indicating management's commitment to the project.

2.2.3 Balancing the team size

An effective concept design team is not hierarchical; the members should be regarded as and act as equals and communicate directly with one another. The topics of discussion often include semi-formulated ideas, and more effort is needed to understand them than is involved in simply decoding the words. A mutual, shared basis for understanding has to be established and developed throughout the project as the ideas progress. And although shared practices decrease the effort required, a separate foundation needs to be laid for each of the mutual interpersonal relationships. The number of interpersonal relationships (N) in a team depends on the number of team members (n) according to the following formula $N=n(n-1)/2$. For a team of 4 members this gives 6 interpersonal relationships, for 5 members there are 10 relationships, and in a team of 8 members there are as many as 28 interpersonal relationships (see Figure 2.2). The likelihood of friction occurring during communications obviously grows with the number of interpersonal relationships. Clearly this friction can be reduced, but a shared ground for understanding cannot be established and developed without effort. At some stage as the team grows bigger, the team building efforts go beyond reasonable bounds. When this happens the team starts to lose its dynamics. Some members' contributions are no longer used or the team may start to work as a hierarchical organisation where communication only follows certain established channels – leaders become managers and partners become suppliers.

Our experience and the above discussions about the necessary roles suggest that a team can easily grow in size. In a larger team, it is more difficult to agree on schedules, decision-making becomes more complicated and it takes longer to get on the same wavelength. In simple terms, the decision-making process becomes slower and less efficient. Therefore, bringing together a concepting team is a compromise between expertise and efficiency. For these practical reasons, the responsibility within a concepting team often lies with a core team consisting of only a few people. This core team is then supplemented with the necessary experts as the work progresses. Lindholm and Keinonen[14] suggest that five members is optimal for a user-interface concepting team to make rapid progress.

A notable exception to the objective of making teams as small and efficient as possible is when learning cooperation and gaining the commitment of different stakeholders in an organisation is an important goal, as in the Ed-design and Decathlon cases (see Chapter 3).

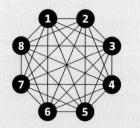
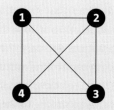

FIGURE 2.2.
Communication in teams of eight and four members

2.3   Briefing the team

A concept design team operating at the fuzzy front end of product and business development is a difficult unit to manage and control. The work typically starts with a loose brief and will end by producing something about which very little was known at the beginning. The team defines the goals, reformulates the questions and adjusts the methods and processes, depending on how the work proceeds. There are of course intermediate results that can be presented for review, but using those for project control purposes can be problematic, because their relationship to the original starting point and the final goal of the project may be redefined during the process. The nature of design problem-solving and the needs of control and management do not make a good combination.

A project plan, which is often obviously used as a planning, project management and control tool, in certain respects does not match well with the way in which concept design teams work. A plan gives the management and customers of the project confidence and security by defining how the project resources will be used and what can be expected as the output. For the team, the plan is in the best case a guess about the future of the project that gives directions and guidance, but in the worst case it is a promise and a chain that restricts the reframing of the goals on the basis of what has been learned earlier in the process. A plan created before the project has really started can hardly be expected to anticipate the challenges and opportunities that the work reveals. Sticking to the plan when the opportunities lie elsewhere is obviously a mistake. When project plans are reviewed from the control perspective outlined above, it is clear that a loose framework and flexible process models are needed to give the work some structure and to allow the resources and schedules to be managed effectively. These issues are discussed in Chapter 3.

Creativity in concept design requires the designer to move away from the design objective rather than focusing directly on the solution. It is impossible to produce new solutions unless the design objective can be seen in a new light. This in turn may require a comprehensive change of perspective. A metaphor frequently used is to take a step backwards to see the bigger picture before rushing forwards. Concepting also includes, for instance, the identification of user needs, technical factors, social values, company goals, the designer's own opinions, experiences, impressions and feelings, and the diverse requirements of the operating environment. Taking these into consideration requires a broader perspective, as well as the use of different working methods and forms of expression in addition to the most direct and obvious ones.

In product development concepting (see Chapter 1), the briefing problems outlined above are perhaps slightly exaggerated. Product development concepts are created as part of a larger product development initiative, and the objectives and restrictions of the project also reflect on the concepting. This is why the factors that apply to briefing design assignments in general also apply to concepting for product development. A fresh introduction to the topic is provided by Peter L. Phillips[15]. However, the challenges of briefing design teams for creating emerging and vision concepts still remain. How should teams that are expected to explore the unknown rather than achieve identifiable goals be briefed?

It is possible, though risky, to start a concept design project without a well-defined plan. Companies that are looking for new opportunities in unfamiliar areas that can turn out to be of significant strategic importance naturally should think twice before taking the chance of simply authorising a concept design team to do what they find interesting. Even in the most open design assignments there are factors that can be used to reveal the starting points for concepting and the process itself, to make it into a more understandable and less fuzzy activity. These include:

- Business strategy and the generic objectives of concepting
- Shared, motivating vision
- Joint management practices
- Trust

2.3.1 Business strategy and the generic objectives of concepting

The management decision that precedes the start of a concept design project is based on clear reasoning. The managers probably have some expectations about the results of the project, but they certainly know why they want to expend resources on something as risky as concept creation. There are always alternative ways to spend the money: launching another advertising campaign, replacing some of the machinery on the production line, etc. They need to justify the choice to their own superiors or shareholders. The reasons they apply are most likely to relate to long- or medium-term business objectives and strategies the company has or is seriously considering. These will obviously also help to keep the team on the relevant track.

Another approach to defining the goals of concept creation is to think about the types of benefits that concepting are expected to generate. Here the generic goals of concept design as identified in Chapter 1 can be used as a checklist. If the concept project is expected to fill an idea bank and file several patent applications, the process will probably differ from that of a project whose main goal is to generate new competences by learning a new technology or initiating cooperation with a recently acquired overseas division that is joining the company.

On the basis of our experience at UIAH, a dual agenda is not uncommon when concepting involves both private companies and public institutions. While the academic partner is working towards a far-reaching vision concept, the business partner may have more immediate expectations and work in parallel to achieve its short-term implementation aims.

In conclusion, the role of the concepting effort with reference to the company's business strategy and the types of goals expected from the project needs to be clarified at the start of the project.

2.3.2   Vision as a starting point for concepting

The initiative for starting a concepting project can be the need to fill a gaping hole in the product portfolio identified during a strategy review, to create a new solution made possible by technical progress or to fulfil identified user needs. However, concept design does not always start with a thorough analysis, nor is it always based on an undisputed scientific finding. The starting point can instead be an attractive idea or unclear opportunity, which offers grounds for further elaboration. It is easy to obtain the support of corporate management for an excellent idea and therefore to acquire the necessary resources for concepting. Typically, when the excellence of the idea is not as obvious, the project needs to be primed more carefully into order to get it off the ground. It must be possible to crystallise and communicate the vision of the concept so that it is easily embraced, stimulating and challenging, even if not much is yet known about its existence. A "soul" must be created for the concept, even though it still lacks a body.

Designing in order to fulfil someone's vision is a conflicting starting point for design. Many designers insist that design is a problem-solving profession where the only correct and acceptable way to start is from a problem statement. The vision-driven approach appears to conflict completely with this view. On the other hand the target-driven and reflective problem-solving style of designers and the poorly defined nature of design problems tend to bring together the problems and solutions into a tightly packaged entity.

Expressing a good vision at the beginning of a project has several advantages. It can give the whole project a direction and goal that is expressed in a challenging way. It states what the project wants to achieve. When the vision is sufficiently open, it leaves room for the team's and the team members' individual interpretations, and thus allows them to make the project a personal challenge. For those people for whom design has to start from a problem statement, an imaginative vision can easily be turned into one. How can this be done? How would it function? What would be the consequences of this type of vision? A vision can also be used to sell the project to other people whose contribution is needed during the course of the project.

The tools that can be used to characterise the concept include metaphors, scenarios and design drivers[14]. A good metaphor describes an idea in a few words, is easy to remember and to pass on, and is widely understandable. A simple and frequently used way to illustrate the nature of the concept idea is to compare the product to be concepted to products in another sector. In conjunction with Nokia's user-interface concepting, mobile phones have been compared to footwear and vehicles[14]. The metaphor of a phone as a rubber boot or an aerobics trainer can easily be deciphered, perhaps not completely unambiguously, but it clearly guides the imagination. Where a metaphor most often refers to the product's physical and design characteristics, a story-like scenario easily illustrates the dimensions of the product's use and interaction. Metaphors and scenarios are stimulating and easily communicated. Design drivers, such as a requirement that the product must be operable with one hand, are more specific definitions of the targets. Drivers are important for putting the vision into specific terms. They specify the problem that will be solved during the concepting process and the information and experts needed to support the work. There can be several drivers which, if necessary, can indicate the different dimensions of the product being concepted. However, paying too much attention to trivial details is not advisable; concepting that starts with a few of the most fundamental objectives is more successful in optimising the major goals.

2.3.3 Ownerships in the concepting process

Phillips[15] strongly stresses the importance of co-ownership in defining a design brief. This also applies to the design of emerging and vision concepts,

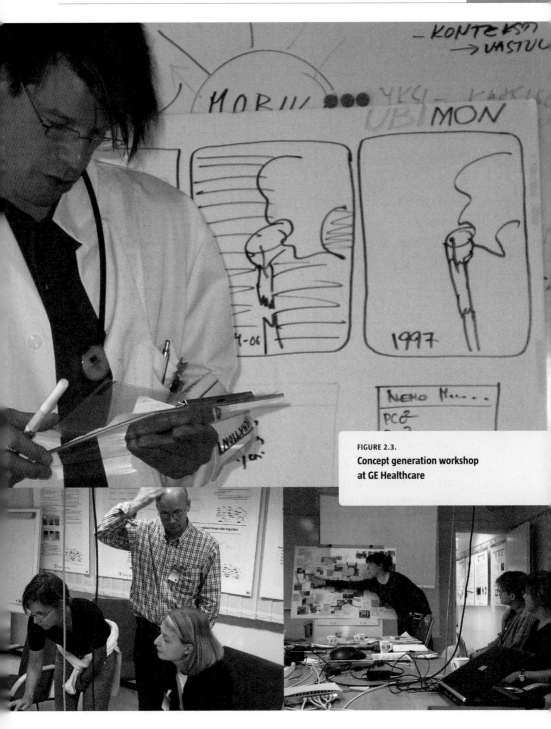

FIGURE 2.3.
Concept generation workshop at GE Healthcare

and perhaps goes even further. The business strategies, generic goals of concepting and driving visions must all be shared by all the key stakeholders. Because exploratory concept design projects should result in the problem statement or vision being readjusted on the basis of the flow or the intermediate results, it is not sufficient to reach an agreement once only. This needs to be negotiated repeatedly, and the communication between the project and the customers should approach a continuous flow of briefing and rebriefing. Obviously, in continuous dialogues the idea of defining a brief becomes vague and starts to be replaced simply by seamless cooperation between the stakeholders. The partner – or the customer in traditional business language – does not leave another partner – the supplier in traditional language – alone with the project, but continues to take an interest and make a contribution after the initiation and between the main checkpoints.

Large and long-term concepting projects often consist of several phases, as many of our examples show, and more often than not the move to a new phase is accompanied by changes in the team. Some of the user experts may leave, and design specialists may join. Some of the designers may leave, and the technology specialists may start their feasibility studies and prototyping activities. In each of these phases someone who knows what has been done leaves, and someone joins who needs to understand how to continue. In each phase there is a need for the previous phase to brief the next one. In these situations, as in the initial briefing, a written report of the results and further actions is hardly sufficient. The briefing needs to take the form of a joint handover period where the premises for further work are interpreted jointly by the people who have created them and the people who will be using them in their future work. (See figure 2.3)

2.3.4   Trust

The need for formal definitions concerning the content and goals of design commissions is twofold: on one hand the understanding about the premises of a project needs to be shared, and on the other hand the formal brief acts as a document to which the partners can refer if future disagreements arise. The shared understanding does not necessarily need to take the form of a written contract, though obviously a written outline of some kind is needed. When the partners have mutual trust, the project does not need to be specified for control purposes. The customer of the concept design knows that the changes the concepting team may make are based on good reasoning. The customer is also aware of the high professional standards and innovativeness of the team. The customer knows that the team is able to produce premium-quality work without the need for a control mechanism. Correspondingly, the supplier trusts the client to have reasonable expectations and to understand if the most challenging aims turn out to be too demanding. Typically the trust is built up over previous joint projects with less freedom and less responsibility for the supplier and with more accurately defined control mechanisms.

2.4   Individual team members

In addition to taking on specific roles, team members also form part of an organisation and are faced with the expectations and requirements of that organisation. Creating something new requires tolerance of uncertainty and the confidence to operate with uncertainties. Challenging the existing solutions also involves challenging the person who presents them. The challenge includes emotional and motivational factors that incorporate the employee's entire identity. Keinonen[16] describes his personal view of concepting as being the pursuit of the limits of credibility. The credibility of design solutions is pivotal in concepting. On the one hand, the team is in pursuit of something that is new and interesting, whilst on the other hand a sense of viability must be maintained. A lack of credibility in the solutions can also mean a lack of credibility in the creator. In concepting, the organisation must allow the individual to become sensitised to open observation and the free presentation of ideas. The support that the team can give to individuals and the team's ability to share the responsibility may turn out to be critical enablers of radically new ideas that challenge established beliefs and practices.

2.5   References

1   **Bessant, J.** (2003): High-Involvement Innovation – Building and Sustaining Competitive Advantage through Continuous Change. John Wiley & Sons, Chichester, UK

2   **Cagan, J. and Vogel, C.M.** (2002): Creating Breakthrough Products – Innovation from Product Planning to Program Approval. Prentice Hall, Upper Saddle River, NJ

3   **Tidd, J., Bessant, J. and Pavitt, K.** (1997): Managing Innovation; Integrating Technical, Market and Organisational Change. John Wiley & Sons, Chichester, UK

4   **Katzenbach, J. & Smith, D.** (1993): The wisdom of teams: creating the high-performance organization. Boston, Mass. Harvard Business School Press

5   **Norman, D.A.** (1998): The Invisible Computer. MIT Press

6   **von Stamm, B.** (2003): Managing Innovation, Design and Creativity. John Wiley & Sons, Chichester, UK

7   **Konkka, K.** (2003): Indian Needs – Cultural End-User Research in Mumbai in Lindholm, C., Keinonen, T. and Kiljander, H. (eds.) Mobile Usability – How Nokia Changed the Face of the Mobile Phone. McGraw-Hill

8   **Kotro, T.** (2005): Hobbyist Knowing in Product Development. Desirable Objects and Passion for Sports in Suunto Corporation. University of Art and Design Helsinki. Doctoral Dissertation

9   **Fujimoto, T.** (1991): Product Integrity and the Role of Designer-as-integrator. Design Management Journal, Spring 29-34

10   **Keinonen, T.** (2003): 1/4 Samara – Hardware Prototyping a Driving Simulator Environment in Lindholm, C., Keinonen, T. and Kiljander, H. (eds.) Mobile Usability – How Nokia Changed the Face of the Mobile Phone. McGraw-Hill

11   **Nielsen, J.** (1994): Usability Engineering. Morgan Kaufmann, San Francisco

12   **Bayer, H. and Holtzblatt, K.** (1998): Contextual Design: A Customer-Centered Approach to Systems Designs. Morgan Kaufmann

13   **Claxton, G.** (1999): Hare Brain, Tortoise Mind: Why Intelligence Increases When You Think Less. HarperPerennial Library

14   **Lindholm, C. and Keinonen, T.** (2003): Managing the Design of User Interfaces in Lindholm, C., Keinonen, T. and Kiljander, H. (eds.) Mobile Usability – How Nokia Changed the Face of the Mobile Phone. McGraw-Hill

15   **Phillips, P. L.** (2004): Creating the Perfect Design Brief – How to Manage Design for Strategic Advantage. Allworth Press, New York

16   **Keinonen, T.** (2000): Miten käytettävyys muotoillaan? [How to design for usability?] University of Art and Design in Helsinki publications B61, Helsinki (In Finnish)

3

Processes of Product Concepting

Roope Takala, Turkka Keinonen, Jussi Mantere

3 Processes of Product Concepting

Roope Takala, Turkka Keinonen, Jussi Mantere

3.1   Introduction

Chapter 1 contrasts concepting with product design. One of the differences noted was the lack of some of the typical product design characteristics in concepting projects. Concept design activities are not necessarily as closely bound to schedules with defined deliverables, product distribution and sales channels or technical and economic constraints as is product design. The expectations for concepting are higher in terms of the creativity, freedom and obligation to innovate and explore. An ideal process for conceptual design would therefore be to allow talented, well-educated, motivated and hard-working people to explore the design space. However, the present trend is for increasingly collectivity. Concept design is seen as a cross-disciplinary activity involving several professionals, representatives of the users and stakeholders belonging to the broad audience of new technologies. The cooperation, if nothing else, sets pressures on managing, facilitating and controlling the creativity processes. Indeed, conceptual design has become an object of scheduling, method development and even computational support[1].

In a business environment, the need to structure the work and to make it more systematic is necessary for conserving key resources or, for instance, being able to present the concepts on scheduled events. However, this needs to be done without applying excessive control that would jeopardize the exploratory nature of the process. Traditional process models of product development and design often focus on controlling design for managerial purposes, which with their predefined milestones do not seem to easily apply to the objectives of concepting.

This chapter discusses the processes developed and used for structuring concept design. We first consider generic characteristics of product concepting processes, and then provide some examples of process models used in industry. All of the processes presented provide a slightly different viewpoint on concepting. The first example is a formalization of how some far-reaching research concepts have been developed at the Nokia research center (NRC) using a user-centred approach. The second example introduces the Deep Dive process used by IDEO, a well-known design agency, for concept design assignments. The last example is the Imaginew process used by Decathlon, a sports goods manufacturer and retailer, to create ideas for new products.

These examples should be considered as inspirational. In practice, the process structures should be modified to support the objectives of the individual project. The processes and structures proposed in this chapter have been highlighted in order to give concept designers an understanding of the underlying nature of the concept creation workflow. Later chapters describe the nature of the concept creation processes in an established business (chapter 5), a constantly and dynamically changing business and product environments (chapter 6), as well as in the creation of a far-reaching vision concept (chapter 7).

3.2   Three generic design activities with main two phases

Few, if any, universal process models for concept design exist or have been published and, as has been explained, those that exist would be difficult to apply in all cases. However, it is possible to identify a general set of design activities that also concepting processes have to follow to solve design problems and to deliver acceptable results. A generic framework for product development and thus concept creation consists of three layers. It incorporates first the acquisition of knowledge that must be in place to allow, second, the successful development of product concepts and, third, their evaluation. The three layers are presented in Figure 3.1. In addition to the downward flow in the process shown in the figure, the model incorporates feedback loops to reflect the iterative nature of the process.

The layers identified as forming part of a concepting process are information acquisition, concept creation and concept evaluation. The design of a product concept requires sufficient knowledge of customer needs, technology forecasts and the business environment. During the concept-generation phase a rough idea, which can come from the background research or from a previously considered driving vision, is expanded into several solution

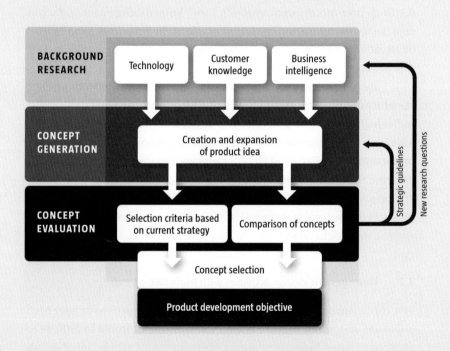

FIGURE 3.1.
The activity layers of product concepting[2]

alternatives, which are then presented for reflection and assessment. The product concepts are evaluated depending on the phase of the project either based on criteria that the presentation itself produces or against a set of fixed criteria, which are often based on a company's business strategy, identified customer needs or business environment drivers.

By their nature, the layers require different working methods. Background research explores a wide range of possibilities to identify opportunities. Concept generation requires the ability to look at the challenge from different angles and to elaborate the initial solutions to produce the concept descriptions for decision-making. On the evaluation layer, the assumptions made in concept creation are reviewed, developed further and perhaps benchmarked against other product concepts and existing products. Finally a decision about the utilisation of the concept is made either as a starting point of a new product or, for example, for communication purposes.

In addition to structuring the concept creation process based on the types of activities described above, it can be divided into subphases on the basis of the object of the design activity. Typically in concept design, and not so typically in product design, the first object of development is a new and different proposal concerning the way the user can work or behave. Based on the results of a user study, or strong insight in some cases, the design team first outlines what the users do, why they do those things, what they expect to get out of the behaviour and where and with whom they do those things. The designers first define a hypothetical user activity[3] only after that they start generating more detailed solutions about the products or services that will enable or support the activity. The Decathlon case below is a good example of this. In some projects, especially technology-driven ones, the new user behaviour may not be the first design result, but may be something that is defined based on the new technical capabilities recognised at the beginning of the project. Nevertheless, the search for radically different novelties cannot take the present consumer needs and user behaviour as a fixed starting point, as this would be much too limiting. New behaviours need to be associated with the new concepts, and because they do not exist yet, they need to be designed and described. Designing on one hand behaviours and on the other hand products and business is a division that can be used in explaining and planning concept design.

3.3   Information-intensive process

Concept design focuses on clarifying the design's input data. This is necessary because only some of the expertise gained from previous products and the organization's previous know-how can be transferred when designing a radically new product and, as mentioned in Chapter 1, the learning itself is often one of the most important reasons for concepting.

A product concept project can be driven by technology whose objective is to find applications for certain new technical capabilities, by business needs in which case-finding concepts for a certain market area is a typical goal or by customer needs when the project tries to find a solution for an identified user problem. Regardless of which drivers are used to initiate the concepting project, a successful process requires input and insight from three sides: realising the technical possibilities, understanding the users' needs and the context of use, and having a working idea about the business models around the concepts. These are shown in the first layer of the model in Figure 3.1. A good analogy for this is given by Donald A Norman who represents a good business case as a three-legged stool, standing on equally long legs of marketing, technology and user experience.[4] (See Chapter 4 for a more detailed discussion of user experience and Chapter 2 for concept design teams.)

An in-depth understanding of all three areas is important for two reasons. First, any one of them can, at any given time, be the driver or the seed for new product concepts. Second, if a company does not have sufficient understanding of them all, this can become a bottleneck in concept generation, evaluation and finally in deciding on the best product concept for further development.

There are numerous cases where a product with superior technology has failed in the market due to inadequate business implementation or market positioning. A classic case is that of the VHS and Betamax video recorder markets. Betamax was clearly technically superior, but the VHS technology took over the market because of a better understanding of customer needs. This took the form of supplying VHS tapes with prerecorded films. Customers did not want to record television programmes – they wanted to watch film when they felt like doing so. In addition, JVC, the company that developed the VHS standard, understood the business concept better than its competitors. Instead of keeping hold of its proprietary technology, JVC licensed the VHS standard rather freely and thus promoted its proliferation over that of Betamax.[2]

FIGURE 3.2.
The Nokia 7280 incorporated a rotating-disc input concept

Technology provides product ideas from the point of view of what can be done, whereas customer knowledge drives product creation on the basis of the needs of the customer and business intelligence from the possibilities of the competitive situation in the market. One example of their interplay is the development of the rotating-disc input device for the Nokia 7280 (pictured in Figure 3.2). The idea for this device arose from the need to scroll up and down long lists of items (such as contacts) quickly, and from the need to differentiate mobile phones. The concept prompted an iteration loop to evaluate several technologies for the input device, as well as development for operating all the devices functions with the new input device. The technology study came up with several options which enabled the device to be technology independent, thereby freeing development effort to focus on usability. An adequate understanding of these issues and the knowledge of the technical implementation eventually enabled the input device to be implemented in the product.

The objective of technology research is to monitor technologies that may be of use in the type of products a company is manufacturing. The target is to identify nascent technologies and understand the product possibilities they may provide. Customer knowledge is a key issue in targeting a product concept. This research is centred on segmentation and on an in-depth understanding of the customer needs that have to be met. In contrast, business intelligence concentrates on the market and the competitive situation. The information allows decision-makers to select market strategies that range from direct confrontation through to avoiding it by a searching for gaps in the market[5]. In essence, research provides a driving vision, a backdrop or a sanity check for product proposals. Research also provides a supply of knowledge that enables fast execution of the next stages of the product development process.

3.4   Concept generation and presentation

A design team needs to come up with ideas and solutions that meet the identified design drivers. It is difficult – if not impossible – to specify a step-by-step process for this activity, where creativity and free exploration play the biggest role. It can be argued that the most important methodological tools for concept creation are those that support creativity and innovation together with those that help to visualise the solutions for designers themselves and the audience of the project.

Three kinds of concepts were introduced in the Chapter 1: product development concepts, emerging concepts and vision concepts. When discussing the methods for creating concepts as a part of the concept design process, the different characteristics and objectives of different concept design cases must be taken into account. Product development concepts aim to define product specifications and often may need quite detailed briefs as input, while emerging concepts and vision concepts can be initiated using targets that are more vague.

The example processes introduced later in this chapter use some common methods for generating concept ideas. In the Decathlon's Imaginew process, brainstorming is used to generate the initial ideas on the basis of background data. Brainstorming is also mentioned in the NRC user-experience design process model as a possible method of generating ideas. On the other hand, designers should be free to use whatever methods they feel will efficiently support the innovation. One example of these is the gadget box used in IDEO's process as described later in this chapter.

Presenting the product concepts commensurably, concisely and clearly is an essential part of concept development. Naturally, the first product concept descriptions are only a rough representation of the final product, but its fundamental characteristics should still be evident. The tools for presenting concepts change along with progress in the project, and mastering the choices of presentation method in every situation is one of the most important process skills in concepting. The concept ideas created using brainstorming, for example, can and should be very brief and general expressed with short phrases or simple sketches. Gradually the selected ideas can be deepened using more descriptive tools. For instance, at first scenario methods can be used. The team starts making stories, cartoon-like storyboards or even video presentations that describe how the users would behave and benefit. Later in the process the team perhaps continues defining the details by creating physical or software-based prototypes. For example, in the Decathlon Imaginew process, scenarios and 3D models of the best ideas generated by brainstorming were created. The way in which concepts are presented is significant, because in addition to conveying the facts it gives the product concept an identity, which is the basis for further development, and enables a general discussion about the subject.

A multidisciplinary evaluation sets its own challenges for the presentation of the concept, because the idioms and ways of describing the special fields will not necessarily be understood by the entire evaluation team. For the more formal evaluation of multiple product concepts, the presentations should be relatively similar in format. This makes comparison easier, and a well-presented concept will not acquire an unjustified advantage solely on the basis of its highly polished presentation material.

One approach in presenting product concepts is to divide the presentation under a few main topics:

1. *Overview of the whole product.* The overview includes the name of the concept, a description of the basic functions and the most important technical attributes. Typically, the physical appearance of the product is also presented in the overview, but sometimes a flow chart of the system can give a better picture of the product.

2. *Definition of the user groups and description of the usage* of the concept product. These descriptions usually contain the market-potential assessment and the usage scenarios presented in the form of narratives or cartoons. If the product targets a user base that differs from the evaluation groups experience and values, the descriptions should receive special attention.

3. *Implementation-related concept evaluation criteria for product development concepts*. Examples of this include the biggest competitors, an estimate of the development budget, an estimate of the time to market and identifying potential bottlenecks related to technical, legal or standards issues.

4. *Test results*. Even if only preliminary tests have been carried out on the concept or its key functions, for example theme interviews or usability tests, the test results are a fundamental part of the presentation of the concept.

Figure 3.3 shows the crystallisation of the DrWhatsOn concept. On the basis of a visioning concept, that focused on the opportunities contextual awareness can bring to the users, DrWhatsOn was developed to include both technological and business-related issues. The final presentation of the concept was crystallized in a one-page description. Additional material included videos on user scenarios, documentation on market potential, costs and other criteria as reference material for the decision-makers.

DrWhatsOn
concept project

CONNECTING PEOPLE IS KNOWING THE CONTEXT

Fingerprint sensor
authentication in e-commerce
applications

Meeting profile changes

Showing recipient context or activity
"Mike is by the sea, running fast"

Context sensitive reminders
"Remember to pick up milk on the way home"

Message delivery based on the context
"Show this to Mike when he goes to lunch"

Receiving of a context profile
"[JFK airport profile received]"

Touch sensor
device in user's hand

Microphones
environmental
audio context

Accelerometers
stability
orientation gestures
user's movements

FIGURE 3.3.

Time and calendar
automated functions
based on calendar
and time information

The presentation of the DrWhatsOn product concept gives an idea of the product concept's basic functions and technologies, and it gives the product concept an identity, name and form

photo: Jari Ijäs, Nokia

Temperature & Humidity sensors
environmental changes

Illumination & color sensors
illuminance
type of light
(artificial or natural)

3.5   Concept evaluation

A survey performed in Finland found that product development personnel often see a new product concept for development just appear out of the blue[2]. The product idea that a decision-maker happens to come across just as product development resources are freed up is the idea that will be developed further. If a company's product development decision-making process is as random as this, it is unlikely that the best ideas will make it to the product development stage, and there is a significant risk that the majority of good product concepts will be missed. Even when a feasible idea finally ends up in further development, it may have taken a long time to establish the development project and to acquire the necessary resources. It is therefore important to pay attention to strategic product decision-making and concept evaluation in order to ensure that the company's best competence capital is used in the products and that the time to market is short.

The evaluation of product concepts is one of the critical steps in the concept development process. The target of the evaluation is to make a decision on whether to discontinue the concept, further iterate the concept or start utilising the concept. Lack of accurate information, in particular for concepts that have longer time scales, often makes evaluation difficult. The use of qualitative evaluation is difficult because the results are not easy to obtain. The use of quantitative evaluation is difficult because of a lack of quantitative data. Issues relating to qualitative and quantitative evaluation are discussed below, together with a list of some of the criteria that need to be considered during concept selection. We also discuss the use of user testing to supplement the evaluation process.

Industrial companies, especially those in the engineering sector, have traditionally relied on various matrix methods when evaluating product designs. Perhaps the one used most often is a scoring method[6], which is also referred to as value analysis. The strength of the method is the clear and unambiguous documentation that shows the evaluation result. In this method the concepts to be compared are given points based on the identified evaluation criteria. The results are charted and a value is given to each product concept by adding up the points for the different criteria. In some versions of the method the evaluation criteria and their importance can first be ranked using weighting coefficients. These evaluation methods are typically at their best when evaluating product concepts that are already a relatively long way into the specification process. When easily comparable quantitative values can be obtained from the indicators, the use of matrix methods is clear and justifiable.

However, there are several reasons why scoring methods are not well suited to early concept evaluations:

1. It is often very difficult to produce quantitative information suitable for matrix methods from product concepts, because important evaluation criteria, such as technical feasibility and the ability to satisfy user needs, can be evaluated only roughly.

2. Matrix methods may often produce only an average solution, because the possible benefits of a unique solution are hidden by the risks. When the objectives of concept design are linked specifically to the identification of new, and interesting possibilities, as is frequently the case, excessive attention must not be paid to the risks even in the evaluation, nor must the benefits be buried under the challenges.

3. One of the weaknesses of matrix methods is the lack of a decision-making rationale. Well-executed product concept evaluations and comparisons, as well as their documentation, give direction to the further development of product concepts.

4. A carefully planned evaluation based on a scoring model may require a relatively comprehensive investigative and measuring phase. In the product concept phase, it is often necessary to minimise the amount of work associated with producing and assessing the evaluation data. This is accentuated if several concepts are being developed in parallel.

3.5.1 Team evaluation

Due to the problems noted above, concepts are typically evaluated using qualitative approaches. When referring to the qualitative evaluation of concepts, expressions such as team evaluation, expert evaluation or heuristic evaluation are used. The expression "team evaluation" is used here in order to underscore the diversity of concepting.

Team evaluations can be supported by checklists, and the evaluations often use prescribed meeting techniques, where the methods improve the quality of the documentation and often help in prioritising product concepts. Typically, the aim of the team evaluation processes is to collect the views of each individual expert and the opinion shared by the entire team. After the experts have become familiar with the concepts at an individual level, the topics they then raise are discussed within the team. The discussion can be used to give the problems or opportunities that are identified a broader context, to ensure their significance and to brainstorm the direction of further development. However, an approach that pursues consensus through informal discussion is probably the most typical way of working.

Although the team evaluation is considerably more informal than quantitative comparison methods, there are a few principles that should be followed:

1. The evaluation should involve experts from a variety of areas so that the study can be carried out in an as professional and wide-ranging way as possible.
2. The criteria needed to evaluate the product concepts should be available to everyone and should be easy to understand.
3. The criteria and the process must be inspiring and stimulating so as to make them suitable for concept comparison, but they must not dismiss promising but unfinished ideas.
4. Concepts must be presented in a way that allows for a multifaceted evaluation that focuses on fundamentals.

At best, a team evaluation is as good as the evaluators, so the evaluation team should contain multiple areas of expertise and backgrounds. The evaluation phases within companies typically involve the best experts from different areas and the people responsible for financing further development. The evaluators must have the support of the bodies granting the resources (if they are not participants themselves).

When checklists are not used in an expert evaluation, the evaluation criteria are considered to be defined by the areas of expertise of the people invited to participate and the bodies they represent. For example, the participation of a usability expert in the evaluation ensures that the usability perspective is taken into consideration – though it does not guarantee that the solution would be usable. This is also one of the weaknesses of the method. When the evaluation becomes dependent on individuals, it is increasingly sensitive to the expertise and availability of the participants. Frequently the most appropriate individuals for the evaluation session are very busy, and poor cooperation skills in the evaluation session can also reduce the value of the results. An experienced facilitator in the evaluation sessions can help to avoid this.

The evaluation of concepts does not always need a comparison of multiple concepts or a selection decision. The evaluation of a single product concept is justified in order to ensure the quality of the concept and to allow it to be comprehensively reviewed. Therefore we can refer to a diagnostic evaluation based on the details of the product concept. The evaluation is then part of the iteration that aims to ensure high-quality feasible product concepts. Possibilities for further developing the concepts and bringing together interesting product characteristics frequently emerge during these evaluations.

A by-product of the iterative team evaluation is the commitment of the participants to the concept project. The people who have participated in the evaluation and whose development suggestions have contributed to the formation of the concept consider the project their own and are more willing to work on its behalf in the future. This facilitates the transformation of concept ideas into products.

3.5.2   Concept evaluation with users

In addition to the internal evaluation methods of companies, as described above, valuable insight into refining the concepts can be gained using evaluation methods involving end-users. The process of evaluating concepts with end-users can have a wide range of different objectives. One typical goal is to identify whether the new product concepts find acceptance amongst the intended target user group. Another goal may be to evaluate the design from a human factors perspective. User evaluation is useful as a tool for iteratively refining the designs based on user feedback in accordance with the concept of user-centred design, as described in the ISO 13407 standard (see Chapter 4).

Market research methods are sometimes used to evaluate user acceptance. A widely used method is the focus group, in which typically 6 to 10 people discuss a topic presented by a moderator, starting with a broader view and then focusing on the specific subject of study[7]. The focus group study as a method of product concept evaluation was recently applied in a concept development project at the NRC. The project had produced a broad idea of a product concept, but no tangible designs. In this project, four focus groups consisting of four to six participants each were used. The studies provided massive amounts of qualitative data to help the researchers identify possible use contexts and user attitudes towards the concept. The experience of the researchers supports Kotler's notion that the greatest challenge of the focus group method is the analysis of the data. The study should be used to gather insights, but not to come to high-level conclusions or to generalise the results to cover the whole user segment[8].

In some cases, more traditional usability engineering methods may be appropriate to evaluate the ergonomics and usability of a product concept during the concept design process. A task-based usability test can be performed on a mock-up, or simulation of the product concept that is sufficiently mature to demonstrate the way it is intended to be operated. In this type of a test 5 to 10 representatives of the intended user group are given typical usage tasks to perform on the mock-up and their performance and comments are analysed to identify areas that need further development.

User evaluations can provide valuable information on the feasibility and details of new product features, especially when the developers are not especially familiar with the user groups needs and habits.

3.5.3 Concept evaluation criteria

It is difficult to define a set list of specific criteria for the evaluation of concepts since it brings out new possibilities and perspectives, thereby partially redefining the evaluation criteria. Each concept can be actually seen as raising a set of criteria particular to it. If a design problem about carrying a device is solved with a neck strap, the safety and comfort of the strap becomes an issue. Another solution to the same problem might bring up considerations about whether a belt clip would be handy enough during driving or whether the target customer segment wears belts. At a generic level, some criteria applicable to both can be found, but the decisions also require deeper understanding and this leads to prioritising particular criteria on a case-by-case basis.

However, at a general level criteria give a framework for concept evaluation. At a generic level it is possible to outline criteria and indicators suitable for product concepting by reviewing the information requirements of corporate management when selecting the product concept. Figure 3.4 lists the factors that affect the selection of the product concept. These factors are based on a survey of industry managers and researchers in the sector. We recommend paying close attention to each group of criteria as a part of the concept evaluation, even though all the listed criteria might not be applicable to all concepts. In a multidisciplinary evaluation, it is also

Product	Fundamental characteristics and features
	Usability
	Reliability
	Environmental and safety aspects
	Product architecture
	Sales arguments
	Industrial design
Technology	Technical feasibility
Customer	User needs to be fulfilled
	Hidden user needs to be fulfilled
	Complexity of the product
Markets	Customer specification
	Market potential
	Time to market
	Supply channels
	Competitors
	Scope of product availability
Profitability	Profit per customer
	Time to profit
	Cost structure of product
Organisation's capabilities	Organisation's realisation ability
	Existing competence
	Extent of outsourcing
	Project leader
Strategy	Product's strategic suitability
	Product risks
	Reverse compatibility
Compliance with regulations and protection	Patent status
	Product-related regulations
	Product-related standards

FIGURE 3.4.
Criteria for measuring and evaluating product concepts[2]

advisable to discuss the content of the evaluation criteria because different people often apply different meanings to the same criteria. The criteria in the figure are by no means conclusive or complete; they should be adapted to suit the type of industry and product in question.

3.6 Iterations and continuity

One purpose of concepting is to identify surprising opportunities. The process used has to be able to adapt and to take advantage of these opportunities. This means that after or even during each phase the design team must decide:

- Whether there is a need to iterate. That is whether to go back through the process and repeat some of the steps with more focused objectives or using modified methods, or whether sufficient effort has been made and there is a sufficient understanding for the project continue.
- Whether to change the approach to the problem.
- Whether the original goal of the project is still valid, or whether there is a reason for reconsidering the objectives on the basis of the results obtained and of an improved understanding of the topic.
- Whether to continue with the original plan (if there is one), for the subsequent stages or to choose the most appropriate alternative methods.
- Whether there is a reason for terminating the project, for example if the opportunities that have been identified are too few and too weak.

It is important to understand how the assessments provide a feedback loop to the other design activities – information acquisition and solution generation. This may give rise to new research questions or a requirement for more iterative concept design loops. It can also lead to a decision to stop iterating and to start developing a product based on the concept. The feedback may also spark new areas of research, seed ideas and guidelines for the generation of new concepts.

As a result of the typically very vaguely defined problems, the iterations in concepting are probably more frequent and the number of iterative loops and the amount of progress achieved by one trial needs to be greater than in product design, where more is known about the end result of the project at the beginning. In concepting there is a lot to be learned, and iterating new hypothetical solutions and then evaluating them can achieve this.

For example, information acquisition on Monday and Tuesday, solution generation on Wednesday and Thursday and evaluation on Friday, and the same agenda for the next week could be the agenda for iterative concepting, but in practice it does not work like that and, even if it did, this would not reflect that iteration defines design phenomena on several different levels simultaneously. It provides a structure for an individual designer's problem-solving efforts whilst trying to make sense of the detail. The iteration can take place in seconds while the eye searches for ideas in collages describing users, hand-drawn sketches and solutions on paper, and the brain assesses the results and adjusts the idea that is in the process of being defined. The more frequent the iteration is and the more tightly the generic design activities are intertwined, the more reflective the process becomes. The generation of ideas can no longer be separated from evaluation. Instead the two activities take place simultaneously while a designer or an effective core team is working. In some cases the layers of design activities are perhaps better understood as parallel threads of activity rather than sequential project phases.

At the same time, in the reflective process a larger loop of iteration can move forward at the project management level. The information-gathering activities and the generation and evaluation of solutions are carried out as separate project tasks or subprojects, possibly by completely different teams that report the results to each other. The extreme situation in which the project results in a product being launched is a single iteration loop (see Chapter 6), where the purpose of the launch is to perform a more reliable evaluation. Our observations suggest that in concept design projects these steps, approaches, periods of intense work and active participants tend to be related to project phases – which we call packages – with a relatively coherent internal structure. In addition, the internal coherence seems to be in contrast to the incoherence across successive packages. Some team members, often a large proportion of them, may change between packages. For example experts only take part in those packages where their special skills are needed. In many projects concept creation is carried out by students who leave the project when a new semester starts. Furthermore, the whole design philosophy on which individual methods are based may change from a research focus to a creative design focus, to fine-tuning and polishing, to an evaluation focus and eventually to a communication and marketing focus.

The distinct packages make project management easy, but they also have their disadvantages. Product design is said to be an extremely hectic activity, but in concept design long periods of inactivity can occur in the middle of the project, often because the hectic product design process consumes all the available resources. Consequently, a common problem in concept design is that the project is split into distinct phases separated by resources, focus and time. The packages diverge and fruitful iteration becomes increasingly difficult. The user research activity simply feeds information into the solution generation process without there being any awareness of the actual information needs, and the evaluation terminates the project instead of nourishing it.

In these conditions ensuring that the project and the information flow freely and fluently is a real challenge for concept creation. The fruitful iteration process can easily be broken up into isolated steps if the packages become too distinct. On the one hand the problem is a project and personnel-management issue involving job rotation or an emphasis on devoting resources to key projects. On the other hand the concepting processes themselves may influence the continuation of the most promising themes.

If the results of the first iteration are robust, relevant and convincing, then they probably speak at least in part for themselves, and will stimulate further development in a second iteration of the concept, but relying on that happening involves huge risks. The most important interim results may not be the most salient. Therefore, there is a need for a means of ensuring the continuity of the key themes across the gaps that occur between the concept design packages. One process-related approach to ensuring continuity is to increase the transparency of the activities. The process documentation could provide access to the interim results and decisions made during the process, which would allow the following phases to gain a better understanding of the results that are passed on to them. However, the process of recording interim steps and results is often seen as burdensome, and is neglected because of a lack of resources. Another approach which has received a more encouraging response is to organise joint handover periods for transferring the contributions to the following package in such a way that the commitment to and ownership of the contributions are transferred at the same time. Practical activities are seen as the most promising way of achieving this. The originators and users of deliverables transform the data into knowledge based on the joint handover elaborations. They create the commitment and ownership for the users and clarify the specific information needs for the originators.

3.7   Timing of concepts

Concepting produces the initial forms of future products. Often the same user needs might lead to different solutions, depending on the kind of technology available. An identified user need or a market opportunity can be transformed into a product on the basis of existing technology but, as new technologies mature, it might be possible to introduce significantly more advanced solutions onto the market within the foreseeable future. In fact, one of the main challenges for concept-related decision-making is to time the process of turning new ideas into products accurately. A vision of how and when customer and user needs will converge with the technological potential must be developed. This vision must be communicated to the correct decision-makers within the company, and the vision must be updated as new information emerges.

One tool used to create this kind of picture of the future is the *road map*. Road maps are strategic documents that companies and organisations use to describe markets, products, technologies, processes and resources in relation to the environment, position and time. Road maps illustrate the relationships between these different factors, they describe future targets and they anticipate the developments ahead. Road maps are used to expand the time perspective beyond the launch of the next product, or beyond the next short-term goal. In this way it is possible to prepare for the changes required by the next generation of products. Continuously updating the road map also helps the company to monitor changes in conditions.

Road maps typically have a horizontal axis for time and a vertical axis for the road map variables, for example the company's product categories (see Figure 3.5). A good road map is clear and easy to understand, and contains only the most important variables and events.

Road maps are used for many different business planning purposes. In this context we are looking at them only from the perspective of decision-making in relation to product concepts, which means that the focus is on product road maps. A product road map functions as the basic document for managing the product portfolio. It shows the company's current products, product families and the forthcoming products that are already destined for market introduction. For each product, the road map shows the schedule for the market launch, how long the product is intended to be on the market, as well as when the product will be replaced and what will replace it. Behind the product road map are road maps describing subsystems, such

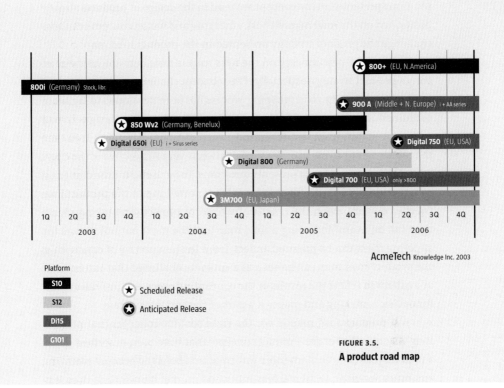

800+ (EU, N.America)

800i (Germany) Stock, libr.

900 A (Middle + N. Europe) i + AA series

850 Wv2 (Germany, Benelux)

Digital 650i (EU) i + Sirus series

Digital 750 (EU, USA)

Digital 800 (Germany)

Digital 700 (EU, USA) only >800

3M700 (EU, Japan)

| 1Q | 2Q | 3Q | 4Q | 1Q | 2Q | 3Q | 4Q | 1Q | 2Q | 3Q | 4Q | 1Q | 2Q | 3Q | 4Q |

2003   2004   2005   2006

AcmeTech Knowledge Inc. 2003

Platform

S10

S12

Di15

G101

★ Scheduled Release

✪ Anticipated Release

FIGURE 3.5.

A product road map

as software and user-segment road maps. Software road maps, for example, show the software applications that form part of different products and how they are evolving.

Road maps have two roles in exploiting concepts. When planning the timetable for the market introduction of the concepts, we recommend evaluating (and, if necessary, creating a road map of) the maturing process of the core technologies required in the concepts, the presumed actions of competitors and the possible market changes. On the other hand, a road map can be used to illustrate the positioning of the preliminary forms of products (which were created as a result of the concepting project) in the company's product categories and the market launch times for these preliminary forms. So, the process of examining the road map and the concept evaluation can lead to a decision to position the concept in the company's product road map either as a totally new product or in such a way that some of the solutions presented in the concept are used in the design of products already positioned on the road map. At best, emerging and vision concepts can also challenge the product strategy presented in the product road map.

Putting a new concept on the road map is a big decision in the company's product strategy, especially if the concept challenges the established product families. The road mapping process has been developed to facilitate coordination, not as a tool for brainstorming. For a concept designer, road maps can, in fact, seem restrictive and even resistant to new ideas. However, the road map approach in concept evaluation is very supportive of the open nature of how concepting presents questions. To be useful for the company, concepts do not have to fill a previously identified gap in the product line. On the contrary, when the results of a concept project start taking shape, they can be examined using a road map and the most natural means for applying them can be pursued. In fact, from the perspective of concepting, the product road map can be seen as a concrete challenge that can be used as a mirror to reflect the results of the concepting process. It provides space for decision-making and makes it a part of the creative process.

A product road map is not the right tool for concepts that probe a long way into the future. Product concepts that have been described using a road map already contain more information about the product platform, customer segment, technical feasibility and market launch time than it is typically possible to specify for vision concepts. However, road map methods can be used when presenting and comparing vision-level concepts. In this case, they cover a longer period in the future and therefore can afford to be less specific.

3.8   Process for user-experience design at the Nokia research center

The ISO 13407 standard for human-centred design[9] suggests that the design process begins by studying background data to understand the context of use and the needs of the customers and/or end-users. The next proposed phase involves generating solutions for the identified needs and use context, then evaluating and refining the solutions iteratively. As a practical implementation, this idea has been refined at NRC to produce a detailed process description that allows the systematic creation of good end-user experiences. Although the process is not an official concept design process at Nokia, which employs several different processes to facilitate its different businesses, it is an example of the way in which some real-life concepts (such as the DrWhatsOn project discussed earlier in this chapter), have been created. The process is aimed in particular at generating emerging concepts with a scope of 3 to 5 years in the future (See Figure 3.6).

The starting point for the process is the user-study phase, in which researchers observe or interview members of the target group. This phase also includes carrying out a background investigation of existing material, other relevant sources of information, as well as market studies and technology prognoses. In the concept creation phase, the goal is to turn the findings of the user study into rough concept ideas. These high-level concepts can be created by brainstorming, for example, and typically describe the identified user need to be addressed, and include a brief explanation of the proposed solution and rough ideas of possibilities for realising the concept. These concepts are then validated to ensure their end-user acceptance and

FIGURE 3.6.
Overview of the User experience design process for NRC

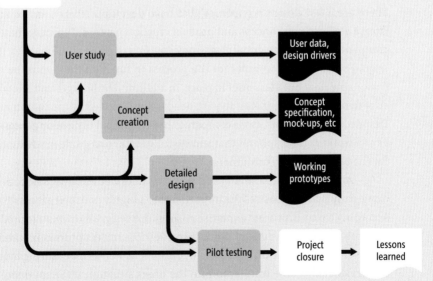

technical feasibility, and to produce a business case. The validated and possibly refined concept ideas are developed into more detailed prototypes and tested for usability, ergonomics and other criteria that are considered important. Depending on the concept and the research focus of the project, the detailed design phase can include, amongst other things, user-interface or interaction design, development of new technical or mechanical solutions or industrial design for the concept.

The process model also proposes pilot testing of the final prototypes of the concepts in a field test before closing the project. This sort of pilot test can provide valuable insights into the how the concept works and is used in its natural use context.

The process model describes the ideal phases that a concept should undergo during its life cycle. It is worth noting that this does not necessarily imply that all the phases should be included in a single project, nor does it imply that the same team should perform all the activities. In some cases, a separate ethnological study project might be run in place of the user-study phase, perhaps ending up with a brainstorming session for high-level concept ideas. Creating a more concrete design on the basis of these ideas might then involve a separate project, possibly with a more technical approach.

3.9   IDEO's Deep Dive process

There are a few design companies that have developed their businesses from a well-defined concept and product creation process. Typically these companies excel in designing clear processes for creative tasks. However, it should be noted that in order for the process to function there must be a reasonably clear business brief in place. In addition, the background knowledge described above, at least in part, needs to be supplied by the customer. Despite all these factors, these companies have managed to develop processes and working atmospheres that successfully foster and guide innovation for the benefit of their customers.

Amongst the more famous of these companies is IDEO (www.ideo.com), a company that emphasises the importance of a highly polished process[10]. According to IDEO, process expertise ensures the design of innovative products, regardless of the type of product in question. IDEO's process is based on observations of the user and usage contexts, as well as on ensuring that the designers become immersed into the user's situation and experience, together with the rapid 3D visualisation and modelling of ideas. IDEO's work interestingly uses the process perspective whilst maintaining the informality, fun and motivation of the design team. One example of IDEO's methods of fostering creativity is the use of toy or gadget boxes that contain various items from toys through to simple components. These help the designers to find examples for their communication and platforms for ideation.

The process begins with obtaining an understanding of the design challenge, which means defining the project's restrictions and focus together with the customer. These are, of course, subsequently tested and possibly amended in later stages of the design project. In the following phases, the hidden motivational factors of the user are of critical importance. The current activities of users are observed to provide answers to questions such as: What motivates the users? What confuses the users? What do they like? What do they dislike? After gaining a relatively good understanding of the current activities, the next step is to start outlining new solutions within the iterative cycle of visualising, modelling, evaluating and refining. Rapid prototyping techniques can be used several times to improve the ideas that are developed before the solutions are communicated to the customer. (See Figure 3.8)

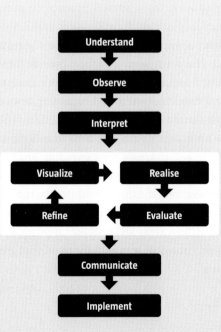

FIGURE 3.7.
The IDEO Deep Dive process

The IDEO process shows that it is possible to provide a service based on the ideation process. There are some modifications to the generic process that adapt the IDEO process to make it easier to use a customer's seed idea and to emphasise the design activities in the process in order to allow for continuous reporting to the customer. The need for customer reporting is also one of the reasons for the rather rigid process structure. IDEO's success, however, does show that structures can be put in place in a successful concept creation process without hindering the creativity of the work.

3.10   Decathlon Imaginew

In contrast to the above case of IDEO, Decathlon is an example of a company that designs concepts primarily for its own use. In the Decathlon case the process consists of a series of phases, together with the best practices and methods that are associated with them. The approach was developed at Decathlon's design center, one of France's biggest design departments, as a process which paved the way for product development. It is used by the design center's advanced design group, which focuses on creating future product concepts and services for new business opportunities and user experiences.

The work is carried out in seven phases: information gathering, brainstorming, scenario creation, concepts, formalisation, evaluation and finally integration with project planning. The composition of the teams involved varies with the project. The permanent members include three designers with unit responsibilities, and three trend forecasters. Designers, design students, product users, sales managers and engineers from outside the company are frequently invited to participate in the different phases of the process. The Imaginew programme primarily aims to anticipate the desires and needs of future users and to create corresponding product proposals based on those wants and needs. The programme enables the iterative evaluation of new product innovations and thus minimises the risk of innovation. An Imaginew programme lasts for about 6 months, and focuses on one of the Decathlon brand's business areas and selected target groups. Below is a case study from the fitness business area with the two target groups "urban girl" and "forever young".

> **PHASE 1 – GATHERING INFORMATION**
> 15 presentations
> 250 presentation pages of background information

In the first phase, the programme's participants gather background data related to their own areas of expertise. In this case study these areas included competitor products, brand innovations, trends, current and future technology enablers and future research, with the data collected by sales, brand management, designers, the R&D team and a specific future research team, respectively. All the teams presented the information they had gathered at a joint meeting. The material was analysed and categorised on the basis of the opportunities and the interests of the team members. As a result the following four themes were identified under fitness: "multitasking", "home integration", "fitness and fun" and "service and advice". (See Figure 3.8)

> **PHASE 2 – BRAINSTORMING**
> 350 ideas
> 21 subcategories
> 4 main themes

A team of about 20 people, representing sales, R&D, product management and the customers, participated in a 1-day brainstorming session to examine the themes that had been identified. The participants presented ideas related to the themes and the target groups. Ideas could be explanations or descriptions of a factor, an event or a new perspective. Ideas were written on notes in a simple format with no detailed specifications. The notes were categorised, and then everyone was responsible for making sure that the ideas remained connected to the brand vision.

PHASE 4 – CONCEPTS
1000 new product and service concepts
200 A3-sized pages of drafts

Next, the scenarios were reviewed to create product and service concept solutions. The team worked outside the normal work environment for 1 week to help them to focus. External designers were also invited to take part in this phase. At the end of the week, the concepts were presented to brand management. (See Figures 3.10 and 3.11)

FIGURE 3.10.
Designing and visualising product concepts

photo: Decathlon

FIGURE 3.11.
Presenting the product and service concepts

photo: Decathlon

PHASE 5 – FORMALISATION
110 product and service concepts
42 of which are relevant from a brand perspective

In the fifth phase, the product concepts were categorised by common traits.
The advanced design group developed more specific descriptions and 3D
models of the concepts that best corresponded to the project goals. Storyboards were developed from service ideas. The aim was to simply model the
essence of the idea, with a simple visual language being applied consistently
to all ideas. In this way, Decathlon investigated the global feasibility of the
product concepts. (See Figure 3.12)

FIGURE 3.12.
Modelling the concept drafts into 3D mock-ups
photo: Decathlon

PHASE 6 – EVALUATION

In the sixth phase, the concepts were evaluated by a team consisting of the brand manager, the design director, the innovation director, product managers and the advanced design group. The evaluation focused on the following criteria: Does the concept correspond to the end user's desires? Is the concept practical or useful in its own sports category? Does the concept help to strengthen the brand? Could it expand the customer base? Is the concept favourable in terms of costs and human resources? Does the solution have a "wow" factor?

FIGURE 3.13.
The *Memory chair* is an easy chair that can be set in different positions for exercising at home

photo: Decathlon

PHASE 7 – GOAL
42 concepts
14 of which are project plan concepts
8 of which are vision projects

In the final phase, the concepts selected were divided into two groups: short- and long-term concepts. Short-term concepts met the identified needs of users, corresponded with the expectations of the selected sport and strengthened the brand. The long-term concepts led to vision projects. It was not yet clear whether or not these projects were feasible. All the concepts successfully communicated the brand vision, they would promote brand development and they had a high "wow" factor. Figures 3.13 and 3.14 show vision-level concepts from the fitness project. The images feature exercise equipment for the home environment.

Decathlon's design director Philippe Picaud notes that the advantage of Imaginew is the possibility of temporarily isolating the designers from their own routines, and thereby allowing them to focus their energy on a single subject area from a broad perspective. This enables and encourages the designer to take part in the configuration of the design brief.

FIGURE 3.14.
Intended for small apartments, the climbing and exercise wall takes up only a small amount of room and has a sculptural appeal

photo: Decathlon

3.11   References

1   **Leppälä, K.; Kerttula, M., Salmela, and Tuikka, T.** (2003) Virtual Design of Smart Products. IT-Press

2   **Takala R., and Ekman K.** (2002)., A Typology of Product Concept Creation and Evaluation, Nord Design 2002, Trondheim, Norway.

3   **Tuikka, T.** (2002) Towards computational instruments for collaborating product concept designers. Doctoral dissertation. University of Oulu

4   **Norman, D.** (1998), The invisible computer, MIT Press, Cambridge.

5   **Cusumano M., and Yoffie D.** (1998), Competing on Internet Time,The Free Press, New York.

6   **Smith, P.G., and Reinertsen, D. G.** (1998), Developing Products in Half the Time, New Rules, New Tools. 2nd ed. John Wiley & sons.

7   **Kotler, P.** (2000), Marketing management Millennium edition. Prentice Hall, New Jersey

8   **Ahtinen, Aino, Aaltonen, Vilja-Kaisa ,Impiö, Jussi, Saunamäki, Jarkko** (2004), How to design a near-to-eye display, Advance, Nokia research magazine 2004:no. 3, pp. 22–27

9   **ISO** (1999) ISO 13407 Human-centred design processes for interactive systems. International Standards Organization.

10   **Kelley, T.** (2001), The Art of Innovation: Lessons in Creativity from Ideo, America's Leading Design Firm. Doubleday, New York

4

User Information in Concepting

Vesa Jääskö, Turkka Keinonen

4 User Information in Concepting

Vesa Jääskö, Turkka Keinonen

A user-centred approach has been widely acknowledged to be one of the cornerstones of successful NPD. A product that is useful, usable and desirable for the user needs to fulfil certain requirements. It must have the necessary practical features for completing a set of tasks, it must be easy to operate and support the user's objectives, it must correspond with the user's values and with the environment where it will be located, and it may even be part of the message that the user wants to convey about his or her personality, by emphasising a certain status, ideological standpoint or professional image. In addition to all this, the product has to stand out from the competition and it needs to be able to convince and attract possible customers at the point of sale.

Designers can make use of their own experiences and visions during the process of designing a new product. This may be sufficient when designing products that come close to the designer's own experiences or when the new product is similar to a previous solution. However, even in these cases, a closer examination of how the product is used and what it means to its owner can elicit previously unidentified facts that can help to identify new product ideas or improvements. The investigations become more important as the complexity of the product increases, and when the designers and their organisation are not familiar with the use of the product. If the probable user of the product is an 85-year-old senior citizen, a research scientist doing chemical research in a very specialised laboratory environment or if the task is to develop a new product and service applications for an extreme skier who spends half the year in the Alps, the everyday experience of a product development engineer or designer is insufficient.

User-centred design has been defined as an activity in which the actual user participates in the design process right from the beginning[1]. Typically the users' activities are examined in the physical and social environments for which the product is intended, and then this information is applied to the development of a concept. Designs driven by user information must participate in an iterative process in which the goals are adjusted on the basis of the user feedback. When it is used skilfully, user information functions as a key source of design inspiration. Karen Holtzblatt, an advocate of user-centred design and a co-developer of Contextual Design[2], notes that a genuine user-centred approach lays the foundation for the creation of pragmatic product innovations that are meaningful to the end-user. At best, including users adds new dimensions to NPD. Working with users commits designers to solving their problems in a completely different manner than if they were simply reading the predigested results of outsourced user studies. Designers will empathise with the users and regard problem-solving as a personal challenge. This can boost the motivation of the individual designers and the team[2].

User information is one subarea of the design data, whose value is determined by how efficiently it can be used in actual product concepting. Therefore, this chapter examines the phenomena related to the use of user information for concept creation. The chapter aims to shed light on what the user information consists of in practice, on the respective roles of users and designers with reference to user information and on how it can be easily and efficiently incorporated into the concepting process.

4.1   Frameworks for user information

The user's experience has become one of the central concepts when describing the user's overall relationship to products and services[3,4]. Because the user experience is ultimately personal, multidimensional and context-dependent, it can never be completely explained or engineered. However, from the practical design point of view, user information is the raw material, that has to be understood in order to put in place the preconditions for positive experiences. But in order to obtain this raw material it is essential to know where to start digging and what to search for. A user-centred approach to design is not a new phenomenon, and user-related information has been used in design in many different ways, depending on the design sector and the prevailing trends. The various paradigms of user information include the following factors:

- Ergonomics and usability
- Aesthetics and semantics
- Lifestyles and trends
- Domestication

Since the 1940s, the ergonomic design tradition has been very influential. It aims to match the functional properties of a product with the ability of human beings to work. The focus was initially on physical product properties, anthropometrics data and basic ergonomics. As complex ICT products began to be more commonly used by non-expert users during the 1980s and 1990s, initially the role of cognitive ergonomics and then the more practical usability approach became the key frameworks. Narrowly interpreted, usability focuses on the interactive characteristics of a system and on measurable performance criteria. These criteria include learning to use the product or system, remembering how to use a new function, efficiency, the number of mistakes made and the level of satisfaction experienced[5]. Sometimes the concept of usability is also considered to include a broader examination of the quality factors related to the context of use, while other people regard the use of this contextual and cultural understanding in NPD as being additional to usability. Nevertheless, contextual user studies and, usability testing have almost become standard design practices in ICT in the 2000s and because of the volume of research in ICT, they form the core of the existing user-centred design tools.

The way the product's construction and materials make the product look and feel have traditionally been the responsibility of an industrial designer. The designer's task has been to sense the topical visual phenomena with which the product should be linked or to which it should refer. When compared to ergonomics, for instance, the working methods have been relatively subjective and insight-driven in most cases. Even though user-centred studies have been and are conducted at various design research institutes to assess and classify the look of products[6] and to analyse the design language[7], the approaches have had little impact on everyday design practice. However, visual characteristics have been designed on the basis of user information in Japan, for example using the Kansei engineering method[8]. This method aims to turn the target segment's verbally specified preferences into design characteristics. Product semantics is another research approach, which focuses more on communicational characteristics and the significance of a product than on its practical operation[9], interpreting design as a kind of language of forms. A conceptual foundation for analysing appearance can be built on the basis of product semantics, but it is not a genuine method of transferring user information into the design.

Studies have also been conducted on the emotional reactions triggered by the product itself and by the use and ownership of it. The users' assessments can be based on the reason for using the product, on previous experience with and knowledge about the product and the usage situation, or it can be based solely on an aesthetic evaluation. According to some authors, pleasure generated by using and possessing products is related to the way in which the product conveys the personality and social status of the owner. The user's set of values and ideology creates the backdrop against which the level of pleasure is defined, whether it is related to the product's design, materials or functional solution. This perspective emphasises the importance of lifestyle and values as a motive behind the purchase decision and a formulator of the user experience[10]. Consumers can make an ideological statement about issues such as environmental friendliness and sustainability[11]. A product choice can also be a reaction against a company[12]. It is worth noting that users compare products with other products and that their choices are based on the existing product offering and on the information available about the product and the company that manufactures it.

Once products reach the environment where they will be used, they can be seen as competing for space in a way that is comparable to how different species compete in nature[13]. To understand the success factors of a product that has yet to be developed, it is important to look at the range of products already being used in households where the new product will be introduced. For example, it is essential to understand which products need or support each other in a symbiotic manner in order to provide a meaningful user experience and to understand which new products, in contrast, will replace existing ones. The way in which the product is used and the space and time available, as well as the roles of the owners and users will all influence these processes. Even though this research interest is often referred to as domestication, the corresponding phenomena also take place in offices and other non-domestic environments. (See Figure 4.1)

According to a model derived from the action theory, users' needs at the level of motivations and actions form two clear and distinct dimensions of user experience that should be approached in different ways[14]. The factors motivating the user's actions must first be understood before embarking on the development of the functional performance that supports the user's goals. In fact, identifying motivational factors is critically important when concepting new product categories, because the new solutions may completely reconceptualise the behaviour, while the users' goals and objectives are more likely to remain unchanged and therefore form a more durable basis for new designs.

The different types of user information have different weightings depending on the product category. With professional goods, the factors related to product ownership, such as social status and the image that the user conveys with the product, do not necessarily play as important a role as they do in the case of clothes and consumer electronics, for example. With an interactive product, understanding the situations and phases of use is naturally a more extensive task than it is, for instance, when concepting new furniture. In addition to the recognised and at least partly controllable factors there can be several phenomena that are indirectly influential, and accessing these or influencing them through design can be difficult. For personal products and gifts, such as jewellery, the significance experienced by the user is essentially influenced by an event associated with the product. Therefore, it is unlikely that any of the frameworks described above would be able to anticipate the experience.

The fact that different product development related corporate functions and departments frequently operate independently presents a challenge for concepting. Industrial designers are responsible for the product's aesthetic and haptic characteristics, user-interface and software engineers implement the product's interactive characteristics, mechanical engineers define the product's physical configuration and product managers in the marketing department are in charge of specifying the product's market positioning. This sort of division of labour can work in a product design project,

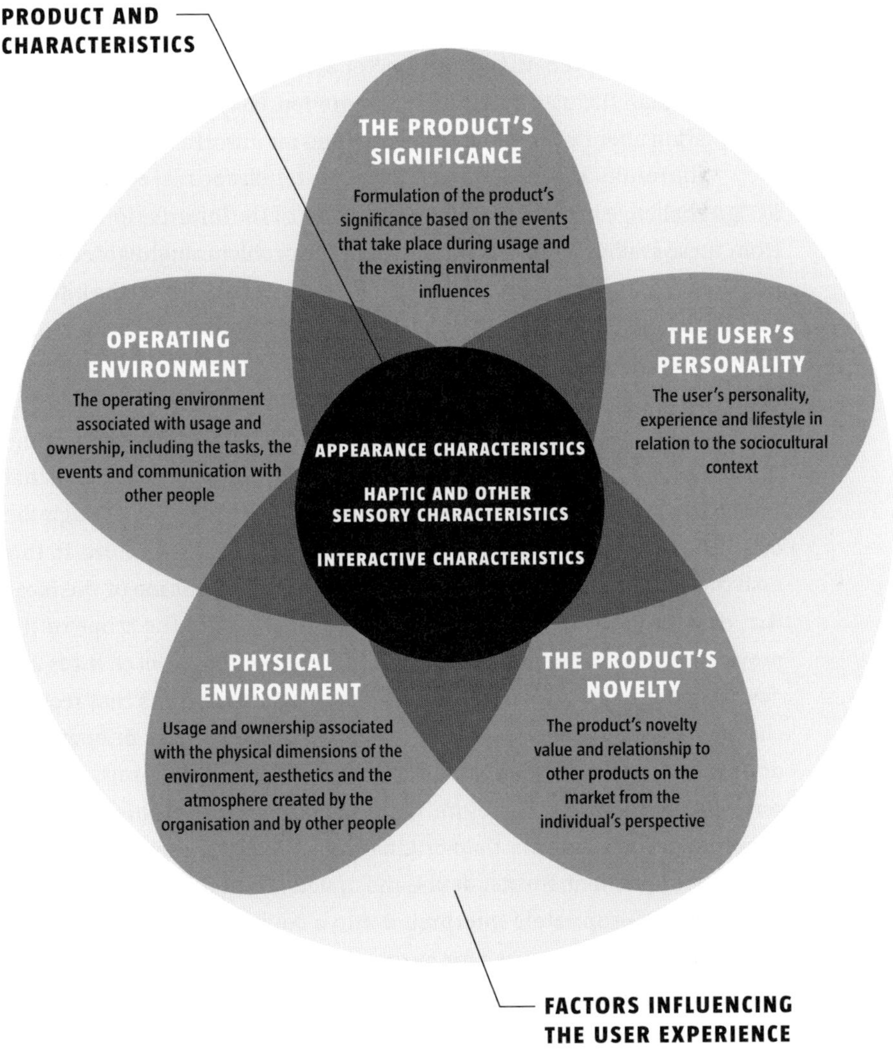

PRODUCT AND
CHARACTERISTICS

THE PRODUCT'S SIGNIFICANCE
Formulation of the product's significance based on the events that take place during usage and the existing environmental influences

OPERATING ENVIRONMENT
The operating environment associated with usage and ownership, including the tasks, the events and communication with other people

THE USER'S PERSONALITY
The user's personality, experience and lifestyle in relation to the sociocultural context

APPEARANCE CHARACTERISTICS

HAPTIC AND OTHER SENSORY CHARACTERISTICS

INTERACTIVE CHARACTERISTICS

PHYSICAL ENVIRONMENT
Usage and ownership associated with the physical dimensions of the environment, aesthetics and the atmosphere created by the organisation and by other people

THE PRODUCT'S NOVELTY
The product's novelty value and relationship to other products on the market from the individual's perspective

FACTORS INFLUENCING
THE USER EXPERIENCE

FIGURE 4.1.
User experience factors in product concepting. User information includes several perspectives such as ergonomics, usability, the emphasis on aesthetic characteristics and on social factors related to the possession and use of the product

although it is not recommended for use in that context either, because the common goal has already been identified at the beginning of the project, can be shared between different units and can be achieved with determined management. In concept design the goals are still being outlined, so setting up cooperation in order to develop an understanding about the user is more challenging. Comprehensive, multisector activities are needed in order to be able to plan the user experience comprehensively. (Chapter 2 provides information about concept design teams and teamwork.)

The results of user studies are usually carefully focused on the concept design challenge in question. As a result, most of the information collected from these studies is specifically aimed at the problems highlighted in the individual concepting project. However, there are ways in which the information can be used beyond the immediate needs of a project. The information collected may include user or environment descriptions or clarifications of operational phases that are generic and are also applicable to other product development projects in the company's line of business. Sometimes the process of observing users may reveal product improvement ideas that can be applied directly to subsequent product versions, even though the original focus was to develop concepts for the more distant future. In this case the information can be used earlier than planned. Some of the ideas that emerge might be impossible to implement within the scope of the project. An improvement requiring multifaceted system-level changes, the development of a completely new product platform or ideas that require changes in the organisation's existing and established ways of operating or in the cooperation between different product manufacturers may be very difficult to realise. Therefore, ideas and information of this kind must be recorded and used later than originally planned. At a later date it is also easier to supplement, further define and update the user information, once it has been appropriately interpreted into a communicable format. User information can also be used for purposes other than direct application to product development. Models describing user actions can be very effective, for instance, when familiarising new employees with the characteristics and usage of products that are being designed.

4.2   Collecting user information

Four main phases can be identified in user-information-based concepting:

1. Collecting user information
2. Interpreting user information
3. Description of new user behaviour
4. Description of a new concept

These phases run in parallel with generic problem-solving approaches and NPD processes. They are included, for instance, in the contextual design method[2], which is one of the most well-known and comprehensive process models in user-centred design. The aim in the first two phases is to construct an understandable picture of the user's present behaviour. This is used as the basis for ideas for a new kind of user activity and a new product. The third phase aims to describe the user's new activity and the user's motivation for the activity and its benefits. Phase four includes the product description and development based on user feedback on prototypes. The emphasis of the phases can vary depending on the product, the goals of the user experience and the project's other objectives and resources. In the following discussion we focus on the first two phases.

User information that directly supports concepting can be difficult to obtain from reliable secondary sources. Some related results might be published as research reports although, as such, the phrasing of the questions they pose is often based on scientific conclusions and does not correspond to the practical information needs of the design team. Product-specific user information already collected by other companies is rarely available for confidentiality reasons and because it has perhaps never been properly prepared for publishing. General, user-related information about consumer behaviour and decision-making, lifestyle and fashion trends is available from commercial and academic sources. The problem with this is often that it lacks the specificity that the designers need in order to define the new behaviour and the concepts accurately. Information related to various hobbies, sports or important areas of life can also be obtained easily from the media, such as trade journals and the Internet. This material can be very inspiring but is rarely reliable. The problems need to be recognised, but not used as an excuse for familiarising oneself with the secondary user information available. Even with its faults, this information is necessary to frame the subsequent research steps appropriately and economically, which consume larger amounts of resources. The role of the preliminary survey using secondary sources that precedes the implementation of a user study in the field naturally becomes more important when the product and application domain being researched are unfamiliar.

Usually the user studies carried out for concepting purposes are qualitative. Qualitative user studies aim to get close to the users and to tap into the way they think, their values, their physical environment and their actions. Even a few users, if chosen carefully, can give the designers extensive new insights. In fact, it has been found that a sample of six people is sufficient for one user group in a qualitative study, even for the development of extensive, interactive system products[2]. For simpler products, the design problems can be identified by bringing just a few users into the design process. For example, if it becomes clear that one 6-year-old child cannot reach the buttons in the lift, bringing in 49 more 6-year-old children to confirm the observation is unnecessary. Once the qualitative approach has been applied to identifying the useful phenomena for generating new solutions, methods that produce more general and reliable results can be used later if necessary.

Even though concepting aims to discover solutions for future products, the concepting process occurs in the here and now. As a result, user information that is available or can be produced now must be used in planning the future user experience. Factors related to the product being designed are discussed with the users as they are today, and their actions are observed in an existing – real or simulated – user environment. The individuals participating in a user study should typically represent the potential market for the future product. For the purposes of future orientation, however, it is often beneficial if the study also includes lead users who may have already solved problems in a new, proactive manner or who can be expected to represent the opinions that the majority of people will hold in the future[2,15]. Obviously it is important to realise that everything that lead users – who can be rather extreme – do and like, will not necessarily appeal to the majority of people.

This same dilemma caused by studying the future in the present that makes user sampling challenging also affects the choice of the products to be studied. Since the product categories that are to be concepted do not exist during the user study, it must examine the use and ownership of existing products that might shed light on the problems of the concepts under development, at least in some essential respects. Sometimes so-called experience prototypes can be built to project into the future. An extreme approach is to produce a fully functioning product for the market in order to determine consumers' responses and, for example, the service development related to the product (see Chapter 6).

There is a wide spectrum of methods available for user studies. In fact, so many different techniques and methods have been developed that it can be difficult for novice designers to find what they need. The techniques, however, are mostly adaptations or combinations of a few basic methods:

- Interview
- Observation
- Self-documentation

Interviews make it possible to discuss the significant factors related to the products, together with their use and ownership. These factors include, previous experiences, preferences, emotions, personal motivations and evaluations, and opinions concerning social status, which are not necessarily apparent using the observation method. The interview is typically supported by a predefined outline of topics that is used to guide the discussion but which can and must be adjusted if any relevant and unexpected issues arise. Interviews can be modified by conducting them in focus groups to identify the users' shared opinions. They can be made more relevant by bringing products or prototypes into the discussion or by conducting the interviews in the actual usage environment. The time perspective of the discussion can be unravelled by asking the interviewees to tell stories about their previous experiences with the products or activities. Interviewees can be given projective tasks, such as making image and text collages, which can bring new perspectives into the discussion that may be difficult to explain in words.

An interview does not allow for very precise descriptions of the operative actions and phases related to practical product usage. Many routine actions are so familiar and automatic for the interviewees that they are often overlooked, and are therefore very difficult to discuss. Observation, in contrast, allows for a very detailed view of a product's usage phases. The people being observed can also be asked to describe exactly what they are doing as they are doing it, in other words to think aloud, which clarifies the objective of the actions and the information that would otherwise not be obvious. They can be shadowed by following them during their daily routines and observed in different situations. Video recording is a good aid for observation and is especially beneficial in longer term or very rapid activities. The environment around the product naturally forms part of the observation. Consequently, in addition to the more obvious functional issues, the observation can include perspectives related to the user's social interaction and existing product selection.

In self-documentation studies, also referred to as probes or diary studies, users record their activities according to instructions and tasks specified by the researchers. Typically, the recording takes the form of diary entries and photographs. The diary material itself can be used as a source of design inspiration or it can be reviewed in a post-documentation user interview for more reliable interpretation. Self-documentation studies are an effective way of identifying users' perspectives on a topic (which are sometimes surprising), of establishing a natural dialogue between the users and designers and of offering insight into situations in which the observer has difficulty becoming involved[16,17].

In most cases the user-study approaches are adjusted and combined. One example of a mix of approaches is a user study about the freeride skiing culture. The study, which was conducted by UIAH and the sports instrument manufacturer Suunto, focused on mapping the phenomena of the freeride skiing culture and on understanding the attitudes, motivations and activities of the skiers. In the mapping phase of the study, an overall picture was outlined on the basis of expert interviews and reviews of secondary sources. These included Internet sites about freeride skiing, video clips, films and magazines. On the basis of this preliminary phase, 10 themes including safety and riders' choices were identified. The picture was then brought into closer focus in a two-phase user study. In the first phase, the users documented their activities and life around the sport using a map, photography tasks and diaries incorporated in a self-documentation package. The daily questions in the diaries were designed to reveal the perspective of the documentation, and were developed on the basis of the 10 themes identified in the preliminary study. The second phase focused on the freeride skiing experience through observation of the activity on the slopes and participation in the skiers' preparations and unwinding activities. (See Figure 4.2)

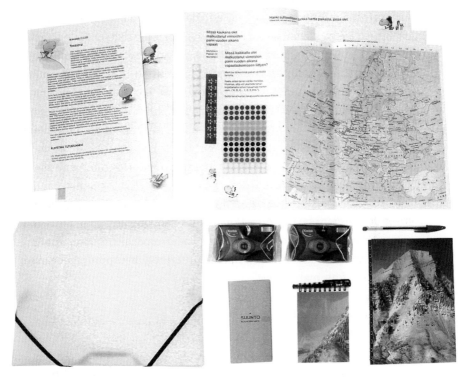

FIGURE 4.2.
Probes package for freeride skiers

4.3 Interpreting user information

When a user study focuses on uncovering potential problems in the use of existing products, it may produce solutions that can be applied directly to the design. In concepting, where the problems and design goals are not as clearly formulated, much of the user information collected cannot be applied in such a direct way to the design. Instead, the information must be categorised and interpreted. It may also be the case that so much material has been collected that it has to be combined in order to be used to support the development work and the decision-making process. Because of the qualitative nature of the user data and, in particular, the exploratory rather than explanatory nature of concepting, rigorous, formal and objective processes to make sense of the information are not always necessary. Therefore, instead of "analysis" we prefer to refer to the "interpretation" of the user data, which leaves more room for insight and empathy.

In the interpretation process, the information is categorised and, where necessary, compiled into descriptions or models that characterise the user, the product usage, the ownership of the product or the environment. These descriptions take information that is otherwise difficult to articulate and put it into a format that can be seen, understood, distributed, assessed and saved for later use. The wide range of interpretation approaches can be divided into four main types:

1. Applying interpretation models
2. Categorisation based on the material
3. Condensing and combining
4. Direct interpretation

4.3.1 Applying interpretation models

Often user data can be organised into specific structures according to a set of recognised and relevant perspectives. The work starts by identifying a point of view that potentially usefully illuminates the material. Next, the organisation principle is applied to the material. The models used in the contextual design[2] method are good examples of categorisation based on previously defined models. The contextual design method has several models that describe the user's activity and environment from different perspectives. The models are used to interpret and model the information collected through observation into a format that supports the generation of ideas for the new solution. The models describe activities by looking closely at the temporal progress (sequence model) of an activity, the communication between people, responsibilities and information transfer practices and vehicles (flow model), the artefacts that support the activities (artifact model), the physical working environment (physical model) and the organisation and atmosphere (cultural model). Other interpretation models include, for example, a cognitive map of the physical environment in the user's mind, a sociogram depicting the user's social network and various descriptions of user profiles.

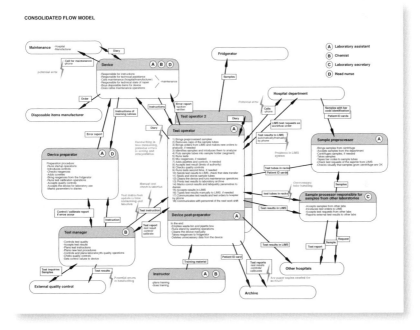

FIGURE 4.3.
A flow model showing the roles, responsibilities and information flows in a work environment

4.3.2 Categorisation based on the material
The disadvantage of interpretation models is that when they are applied, the data only reveals what the model assumes and allows it to reveal. Often, however, one of the most important aims of a user study is to find or create new structures describing the existing practices from fresh angles that support the creation of surprising concept solutions. Therefore it is necessary to

allow the data itself to influence the categories into which it is organised. The interpretation follows a bottom-up principle, which starts with details and moves up to structures on a more general level that describe the data and offer the most useful organisational principle. The most familiar methods based on material categorisation are probably the grouping techniques (e.g. the affinity diagram[2] and language processing[18]), where the significant issues emerging in the user study are first recorded and then categorised by a multidisciplinary team based on their affinity. In addition to textual material, affinity-based grouping can be applied just as effectively to categorise visual materials. Compared to the application of interpretation models, creating new structures based on the material is a much more demanding process and hence requires a competent and experienced team.

A video card game developed by Professor Jacob Buur was applied in the freeride skiing project. The game combines the idea of the happy families card game with the categorisation of video clips. The objective of the card

FIGURE 4.4.
Video cards and the themes of the different phases of freeride skiing

game is to work in teams to categorise short clips taken from video footage so as to give the data a meaningful structure. First each player is assigned several video clips, which they watch and comment on. In the card game, related clips are linked and the linking themes identified. In the freeride skiing study, the previously unidentified themes were primarily related to the reasons behind the phenomena that had been identified, for example "Why does the skier want to radio to the next skier about where to ski down the hill?" On the basis of this new understanding, it was possible to obtain a clearer view of the values behind the skiers' actions and attitudes. (See Figure 4.4)

4.3.3   Condensing and combining

A user survey often produces material that itself essentially illustrates the factors that are significant to the user. However, the problem is that the most impressive items may be concealed by the body of uninteresting data. For example, some photographs taken in conjunction with the observation are very effective in presenting the atmosphere of the user's environment, whilst others just document the shelves and walls. Personality descriptions based on photograph and video material can also work in the same way. The best pictures are worth more than a thousand words, while the rest may be completely uninteresting. The data need to be filtered, with the material that is assumed to be most characteristic and most useful for the development of new ideas being selected. In a teleworking user study by UIAH, the participants and their apartments were photographed in conjunction with interviews. The photographs and descriptions that best illustrated the families were compiled into family-specific collages, with these collages conveying a clear image of the atmosphere and spatial arrangements of the family homes. These are essential in order to understand the physical impact of teleworking on the homes themselves and on the other activities that take place there. The material was presented to designers in a workshop, who used the material to draw and act out scenarios that focused on new ways of working at home. (See Figure 4.5)

4.3.4   Direct interpretation

In direct interpretation, the people who conducted the user study, or those who have reviewed the as yet uncategorised user-study material, brainstorm new product solutions and their usage scenarios directly. The stimuli may

FIGURE 4.5.
Presentation of user information related to teleworking at home

include material such as diaries or photographs taken by the users themselves. Concept brainstorming can also be carried out directly in the product's
usage environment whilst the product or service is actually being used. In
this kind of contextual brainstorming[19] the designer's personal experience
is combined with the team's capability for generating ideas, which can result
in new product improvements and product ideas being developed. These
approaches work best in relatively simple design tasks where the data do
not need to be presented to others and where it does not need to be used
to justify the solutions. On the other hand, this approach can be a useful
alternative when the problems are so complex that intuitive and emotional
decision-making may be more likely to hit the target than formal analyses
which, as the complexity of the problems increases, can themselves become
too complex. (See Figure 4.6)

4.4   Relationship between users and designers

We have already discussed the different traditions of user information that
are applied in NPD and some generic approaches for gathering and making sense of user information. In order to obtain a more comprehensive

FIGURE 4.6.
**Brainstorming of new teleworking
practices using scenarios**

understanding of the variety of current practices, we introduce yet another point of view, namely the nature of the relationships between the users and the designers. The relationships described in metaphorical terms aim to indicate the closeness of the interaction between users and designers, the perceptions about the sources of expertise and creativity in the interaction, as well as the situation and the context sensitivity of the relationship. The relationships we have identified are:

- Engineer designer and component user
- Doctor designer and patient user
- Student designer and master user
- Coach designer and athlete user
- Artist designer and muse user
- Chef designer and customer user
- Director designer and character user

4.4.1   Engineer designer and component user

Designers seldom have the freedom to define their product from scratch. Instead, they have to take into account several kinds of specified requirements and use available components. The system has to be designed so that the components, which have existing technical specifications, fit together such that they enable the system to meet its performance goals. Some of these components are standardised, such as nuts and bolts, and some are chosen from the catalogues of different component suppliers, which allows for more variety. However, the choices are limited and the system architecture is influenced by the dimensions and other characteristics of these components. It is not possible to order tailored components for all the applications – generic solutions have to be used without context-driven optimisation.

The human operator can be seen as one of these components. The dimensions of the human operator can be obtained from human factors manuals, such as the well-known Dreyfuss anthropometric tables[20]. In these manuals human beings are presented as components in a suppliers' catalogue with specific dimensions and performance capabilities with, of course, a certain amount of variation because of the unreliable manufacturing quality of the human component. The average height of component type X, for example European women, can be identified, as can the average power of the upper limbs of component type Y, for example South-East Asian men, or the short-term memory capacity of type Z. This kind of data relating to the human factor, or component, is extremely useful and often provides the designer with the necessary basis for decisions relating to the human operators. The data are generic and not depend on what the components are used for. The grip power of a hand is the same, regardless of whether it is holding a screwdriver or a remote controller. Obviously the component itself has not been asked about its capabilities, but instead its performance has been measured objectively.

4.4.2 Doctor designer and patient user

From our own experience we know that the process of adapting and utilising new technologies in professional activities and in everyday life is not always straightforward. Information technology is difficult to use if the features and user interface do not correspond with the users' activities and mental models. Knowing the users' short-term memory capacity does not help if the designers are not aware of what it is used for. The tasks that the users carry out specify the requirements for the users' capabilities, and the different ways in which the users conceptualise the tasks result in different capabilities being used. Therefore it is not sufficient to be familiar with the generic user; the designer must also be familiar with the different tasks, and the variety of tasks combined with the variety of individual capabilities makes generic modelling too complex. In this situation the focus of attention shifts from the user to the interaction between users and artefacts; it shifts from the generic to the particular. The most practical way of handling specific interaction is to isolate the problems that need to be resolved. "If it works, don't fix it" says the old engineering slogan. The designer, who used to regard the user in the same way as an engineer looking at components in a supplier's catalogue, has now started to look at users as a doctor looks at a patient in the context of clinical medicine.

Fluent interaction is like a healthy body. The problems in interaction can appear, at worst, as symptoms of an illness such as stress, an excessive workload, tiredness and frustration. The task of a researcher/designer/doctor is to diagnose the symptoms, isolate the reasons behind them and cure the interaction. A usability test is the obvious example of the clinic where the symptoms are identified and medication prescribed. Even though modern medicine underlines the importance of good dialogue between the patient and doctor, the roles are still clear: the patient brings in the problem and the doctor applies the expertise to solve it. The same applies to the usability testing paradigm: the user generates problems and the developers use their expertise to solve them.

4.4.3   Student designer and master user

Anticipating and avoiding problems is more efficient than resolving them. In order to be able to do this the designer must be familiar with the specific interaction before it is demonstrated in the research laboratory using functional prototypes. The designer needs to learn about how the specific interactions take place in certain activities. On the basis of this understanding the designer can propose solutions that do not lead to situations where doctors are needed. In contextual observations conducted, for instance, in accordance with the principles of the contextual inquiry[2] method, the designer aims to become familiar with the activity by observing it, in the same way as an apprentice learning from his master. In this relationship, the users are the experts in their fields. They are experts who are respected and whose expertise is not questioned, though the designer may ask a lot of questions in order to understand what is happening. There is a reason for every action taken by the user, and understanding the reason can be important to the design. This approach helps designers to eliminate their own existing assumptions that might prevent them from seeing the true nature of the activity and the factors that are important to the user. In comparison with the previous metaphor, the owner of the expertise has changed. The designer, who used to be an engineer making use of the user and a doctor solving the user's problems, becomes a student who listens to and watches the master-user's behaviour.

4.4.4   Coach designer and athlete user

The designer-students learn from the master users when they visit their premises. However, when they return to their studios the designers are once again the experts and use the understanding they have acquired to generate solutions. In line with the Scandinavian tradition of participatory design, the user should be also able to participate in the development of new solutions. This principle originates from the design of working environments in which the users' influence on the development of their own workspace has been seen as an instrument of workplace democracy[21]. Participatory design builds on the idea that users themselves can articulate and interpret their needs with respect to the functionality and pleasant atmosphere of their own working environment. The concept also has its advocates in user-centred product design. This approach emphasises the communication between users and designers and the equality of opportunity to influence the design. The users, who are considered experts when proposing design solutions concerning their work, need to be given special equipment to deal with design problems, because after all they are not design professionals. Thus, the role of the design professionals is to facilitate the interaction between different stakeholders whose participation is needed, and to give them the means to deal with hypothetical future options. The designer students become designer coaches providing the master users with the set of tools and skills they need to express their ideas. The coaches know the rules of the game and can plan the tactics, but they need the users to perform; the expertise is thus shared between the partners.

4.4.5   Artist designer and muse user

The European commissin funded Presence research project looked at the social activity of people living in residential areas considered to be problematic and their perception of their own environment. The goal was to determine the character and background of these people and what they considered to be important in everyday life. To do this, the researchers developed a documentation tool, a package, that was sent to the participants in the study. The creators of the method call this self-documentation material cultural probes. The package contained, for example, photography tasks, questions and exercises that resulted in the production of various visual materials. The documents produced by the residents in accordance with the package were sent to the developers. The developers were no longer in contact with the people, and they did not even necessarily know who had produced the material[22]. Therefore, the interpretation of the user information was fully the responsibility of the designers. The central goal of the work was not really to obtain accurate, factual, verifiable information, but to produce material that had already been processed by users and that could be used to inspire the creation of surprising product ideas.

This example illustrates yet another different relationship between the users and the designers. The designers are artists looking for inspiration and they have chosen particular users to be their muses and to inspire their creative processes.

4.4.6   Chef designer and customer user

The emphasis of the designers' own, experience-based interpretation of user information is apparent in the working methods of some successful design companies. The idea is that a true understanding can only be gained and used to support the design process after the designer has personally experienced the usage situations and the function of the product that is being designed. IDEO's user-centred design process consists of several phases, the first of which maps and creates a background account. In the ensuing user study, the user's activity is observed and the related factors are discussed with the users. On the basis of this information, the designer can empathise with the situations experienced by the user and can brainstorm solutions. The designers interpret information throughout the process, but it only crystal-lises and becomes visible when the concepts emerge during the empathising and brainstorming phase. The designer can be seen as a chef, concocting new flavours in the kitchen. The chef has to taste the flavours himself in order to be able to develop them. Of course, the customer can suggest the flavours that he likes, but it is not his job or even within his ability to create new flavours[15]. But obviously, the customer user can choose whether or not to come back and eat at the restaurant again.

4.4.7   Director designer and actor user

The step from the designer acting as a user, the chef who trusts his own taste, to designers themselves creating an image of a user is not a big one. Instead of simply creating the solution, they also create both the behaviour and the actors who act out that behaviour. Sometimes the creation of the potential future user can be directly based on the user research, as a result of the consolidation of information, as discussed above. Sometimes design-ers add characteristics that have not actually been observed but which are considered necessary, perhaps to stimulate ideas from extreme users or perhaps to make assumptions about future behaviour that does not yet exist. The designer behaves like a theatre director turning a manuscript into a play – sometimes the director follows the manuscript faithfully, but often the drama calls for changes.

It is possible to criticise the use of user descriptions that are not directly based on user research because they are not user-centred. However, the act of creating them and using them in the design as hypothetical users at least focuses the attention of the design team on different kinds of users, which prevents the designers falling into the common trap of designing for themselves. When a user is created, it is easy to continue by creating the user's behaviour, which is typically presented in design scenarios, that is stories in which the user's new behaviour is illustrated in narrative form.

It is common for the users' and designers' roles to change as the design project progresses. At its best, a successful user-centred design includes all of these roles. It inspires design and creates innovations, but it is also analytical and, from the user's perspective, leads to the correct, tried-and-tested solutions.

We will now look at two user-centred concepting case studies to demonstrate the usefulness of the user data that are collected and their implications for product design. The first case study focuses on the concepting of an interactive product in a large corporation. The second case study is a relatively small-scale concepting project, in which a low-tech, mechanical, muscle-powered vehicle is being developed. The case studies have different concept creation objectives (see Chapter 1), they focus on fundamentally different types of product area and, because of the different development settings, the resources available to them differ substantially.

4.5   Concepting in an in-car communications user interface

A steering wheel user-interface concepting project was conducted in 1999 by NRC and Nokia Mobile Phones (NMP). The participants in the NMP project came from the accessories unit and the unit specialising in factory-installed car phones. Both units needed new products, but the project results were not allocated to any specific product or project. So the design was channelled freely, on the basis of starting points identified by the design and research team, towards products that were to be realised at some unspecified date in the future. The secondary goals were to increase the company's general understanding of user interfaces in the car environment, to create cooperative links with the automotive industry and to protect possible key solutions with patents.

The project examined several issues related to the communication user interface in a car, such as voice support for text input[23,24] and the presentation of visual information to the driver. The project identified, for example, the need to divide the communication system's input and display devices into two separate components. This solution is relatively obvious, but the project focused on the reasons for doing this and the details of the solution. When the company later needed a user interface platform for a new series of car phones, the Nokia 810 (Figure 4.7) introduced in 2003, its ergonomic concept had already been researched, and the design process was able to move quickly on to tackling the concrete issues and specifications.

The steering wheel project also examined and tried to anticipate the evolution of visual design solutions within the car environment in order to ensure the best possible fit of the next series of car phones with the design of the dashboards in new cars. This work involved learning about the automotive industry's newest design concepts and transferring their shapes and visual language to the communication devices. However, here we examine the project from the interaction design perspective, concentrating on the physical and cognitive ergonomics areas of product concepting.

The project followed the principles of user-centred and cross-disciplinary design. User information was collected and processed using several methods, the main ones being the observation of users through contextual inquiry[2], learning about traffic psychology study data, the design team's personal experiences with various vehicle user interfaces and the prototyping of solutions and usability testing with interactive simulations.

FIGURE 4.7.
The Nokia 810 car phone

The objective of observing and interviewing drivers was to identify salient behaviour models related to in-car communications and to understand the reasons behind them. One of the key findings was that when the driver received a call while driving, identification of the caller was necessary in order to make a decision on the action to take. The majority of the drivers observed decided what to do on the basis of who was calling. A brief call with a simple conversation was not seen as disrupting or endangering the driver's performance, but if the call was expected to last a substantial length of time or to be demanding in content, drivers tended to avoid the call. So, in practice, calls were answered if they were from a close colleague or from a family member, because it was felt that a conversation with them did not require too much attention. Drivers did not want to talk with customers while driving, because it compromised their concentration on the call or on their driving. The results were very similar to the results obtained in the controlled tests on the impact of the cognitive demand of a secondary task on the driving performance run by Heikki Summala, a professor of traffic psychology[25]. The actions of the drivers observed in the projects were very rational. Identification of the caller was of primary importance for communication and for smooth and safe driving. The GSM standard supports the identification of the caller, but there was room for improvement in how this information is conveyed to the driver.

Another issue raised by the observations was the unergonomic positioning of the car phones. Mounting the phones on the upper part of the dashboard, where it is easy to read the display, proved to be problematic. Even though mobile phones are relatively small, the strict safety regulations for devices installed in cars require the phone holders to be large. Many car models do not have a suitable location for the convenient installation of large holders. Moreover, car owners want to avoid damaging the visible areas of the dashboard with the holes required for phone installation. Perhaps the most surprising observation was that colourful phone covers distract the driver if the phone is installed close to the central field of vision. A good example of this is the picture taken in conjunction with the observations conducted in London (Figure 4.8), in which a bright yellow phone was installed on the centre console next to the gear lever so that the bright colour would not distract the driver.

The observations of the users highlighted caller identification as one of the key factors affecting communication and driving safety, and that

Vesa Jääskö, Turkka Keinonen

FIGURE 4.8.
A yellow Nokia 8110
mounted on the car's
centre console

the attributes of the phones and their holders often led to unergonomic mounting positions.

The project's understanding of traffic behaviour was supplemented by studying traffic psychology. One of the studies examined dealt with the impact that the placement of a monitored display had on the driver's reaction time when the car ahead started to brake. The results of the study (Figure 4.9) clearly indicated that the correct placement of the display, particularly in terms of its height, is a critical factor. Drivers must be able to see the key information without having to lower their glance below the upper part of the dashboard. This was in clear contrast with how car owners typically install the mobile phone car kits available on the market.

In addition to the location of the display, the correct amount of information and the way in which it is presented are important aspects to consider when designing information display systems for drivers. In 1999, when the project was underway, several car makers were introducing integrated car systems equipped with large displays. Figure 4.10 shows Daimler Chrysler's integrated communication, navigation and audio system user-interface solution designed for the Mercedes Benz S-Class car. The project had an opportunity to test this solution. On the basis of the look and feel, it seemed clear that the large display's ability to present a lot of information could not be used in a driving situation. Moreover, traffic psychology studies indicated that the risk of losing control of the vehicle increases to dangerous levels if the driver's eyes are off the road continuously for more than 1.5 seconds[26]. The necessary information must be found on the display immediately, so the project proposed a small, one- or two-line display.

Using the research work, the project was able to determine, amongst other things:

- On the basis of its own observations, the most essential phases of interaction affecting driving safety and communication flow, and the information the driver needs in those phases.
- Again, on the basis of its own observations, the driver's preferences related to the positioning of the phone.
- On the basis of published sources and the project's own trials, the positioning of the display and the amount of visual information in a vehicle.

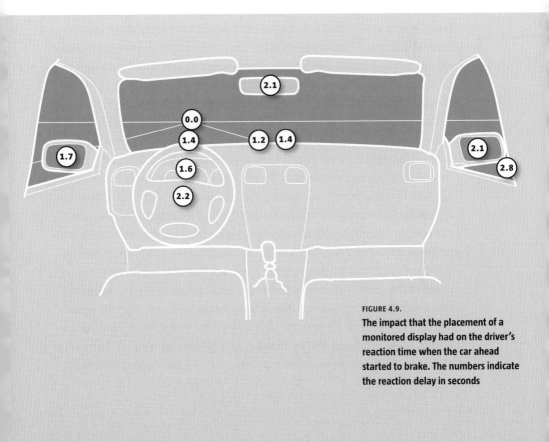

FIGURE 4.9.

The impact that the placement of a monitored display had on the driver's reaction time when the car ahead started to brake. The numbers indicate the reaction delay in seconds

FIGURE 4.10.

**Navigation, audio and communica-
tion user interface of the Mercedes
Benz S-Class car from 1999**

After determining the central design goals, the concept's next design challenge was to review Nokia's user-interfaces and to modify them so as to
accommodate the requirements of the car environment and to ensure that
the solutions would be both compatible and consistent with Nokia's other
products. The user interfaces of Nokia's mobile phones are largely based on
the use of so-called soft keys[27,28]. The display text for these keys changes
to indicate their different functions. Even though viable and commercially
successful user interfaces have been built into mobile phones using soft
keys, using them in a car was seen as problematic. Previous studies have
demonstrated the importance of the close physical proximity between the
soft-key text and the corresponding soft-key. However, in terms of driving
ergonomics, the keys and displays need to be in different places. Moreover,
the correct interpretation of soft keys requires the user to read and com-
prehend at least two display texts: the name of the key and the name of the
menu function that the key targets. It was felt that this would take up too
much of the driver's attention. So the project started developing solutions
that would be as compatible as possible with mobile phones and that would
use soft keys as little as possible. A simple driving simulator was built to
design these solutions and to analyse the designs (Figure 4.11). The user-
interface design solutions were built directly into the driving simulator,

FIGURE 4.11.

**A simulation environment for the design
and testing of a vehicle user interface**

where they were first tested within the design team and then by people outside the design team.

The research and design work described here, which was carried out alongside the other issues examined in the project, took about 9 months. The process that started with the observation of the drivers ended with the simulation tests of the designs that had been produced. A similar design process could have been carried out as part of a product development project. However, it is seldom possible to allocate this amount of time to produce only the user information in conjunction with a product development project. A user-centred approach is possible in a concepting project that is implemented early enough and separately from the product development process. The solution that emerges can be used quickly if it is later decided to launch a product project. In this case, the time between concepting and utilisation was 4 years (1999 to 2003).

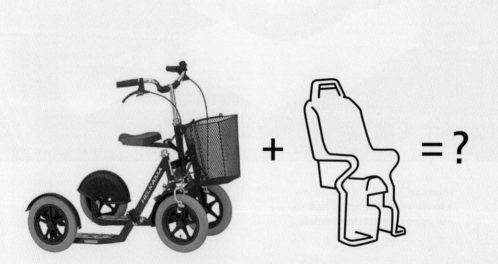

FIGURE 4.12.
Helkama four-wheel kick bike for the elderly and a child bicycle seat
– the starting point for the concept creation

4.6   Four-wheel kick-bike concept development

The following case study describes a concept development project that was implemented by a Finnish design consultancy, Muotohiomo, for a medium-sized Scandinavian bicycle manufacturer, Helkama Velox (later Helkama) during 2003 and 2004. Helkama also produces tricycles and four-wheel kick-bikes for elderly and disabled people.

The case study describes the development of a concept idea in which a child bicycle seat was integrated into a four-wheel kick-bike (Figure 4.12). The aim of the project was to introduce a new category of muscle-powered, short-distance vehicles for families with small children. The opportunity seemed promising enough to merit closer scrutiny, and an industrial design consultant was hired. The basic idea of the concept was outlined at a meeting between the product development manager and the industrial design consultant who formed the core team of the project.

The assumed primary target market for the concept was young families living in suburban neighbourhoods with daily, close-range mobility

positive image

negative image

with 4 wheels = ?

FIGURE 4.13.
The challenges and target of the new concept design

needs. Therefore, the main goal was to develop a new product for carrying a child, but it was also important to be able to produce a version of the same basic product for elderly people without making extensive changes. The dual focus was regarded as an opportunity to bring the use of four-wheel bikes, which have been regarded only as assistive devices for senior citizens with walking problems, into the mainstream. The challenge was to create a sporty design and appearance without any negative associations with walking aids – otherwise the concept design brief was left open so that it could be formalised during the design process. (See Figure 4.13)

During the first concept design phase, six families documented their daily activities relating to travelling with children using their available methods of transport. The documentation period took 4 weeks, with the documentation taking the form of self-photography and a diary. In our experience, self-documentation material is seldom sufficient to give a proper picture of the activity being studied. Therefore, the aim was to acquire material to inspire the design and also to prepare the families for the subsequent interview session. In the interviews the self-documentation material was discussed in order to ensure that it had been correctly interpreted and to gain a deeper understanding of the documented incidents. At the end of the session, the new concept was introduced with illustrations showing the design, product features and scenarios of different use cases. The families also had a chance to test a functional prototype of the new concept that had been developed in parallel with the user research.

The conversations were recorded and annotated to identify relevant remarks, potential problems and design opportunities immediately after the sessions. The aspects that needed to be considered and their relevance were evaluated by the design consultant who was familiar with the information needs of the design process and who was in charge of the design development. This streamlined process, which did not involve transcribing the interviews, is possibly not adequate in scientific terms, but it is an efficient way of drawing conclusions from the data.

Several approaches were applied in parallel because the aim of the study was to cover a variety of different user experiences. The self-documentation included drawing a picture of daily outdoor activities with a child. This was useful because in this particular case it was considered necessary to obtain user data over a relatively long period of time. The process of creating the documentation was guided using questions such as: What are the

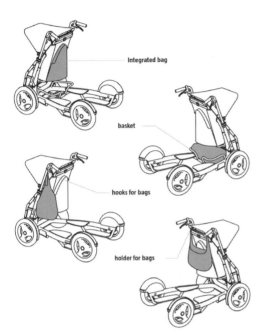

Integrated bag

basket

hooks for bags

holder for bags

↑↓ FIGURE 4.14.

Illustrations from the concept presentation, showing scenarios of carrying different loads and travelling with children of different ages

tandem seat

child´s footboard

different phases involved in going out with a child? What motivates you to go out? Why do you use a car, a bus or a local train? In the interview it was possible to go into more detail and to investigate specific personal issues. New concept presentation and testing revealed the first impressions, the perceptions of new features and the comparisons with existing products. Testing the actual physical product helped the designers to understand the level of acceptance of the new way of travelling and the child's reaction to the vehicle. Another reason for the development and user research being run in parallel was the limited resources. Consulting time was saved by merging user-study interpretations and concept evaluations into one session. (See Figures 4.14 to 4.17)

The outcome of the study was consolidated into the following findings:

- Relevant information for development, grouped as remarks under different user-experience topics
- Features that needed to be developed
- Comparison of the kick-bike with existing and competing products
- Critical success factors

According to the study, the basic concept idea was regarded as mainly positive. The fitness image seemed appropriate for the product and could motivate the parents to use it, which is an important issue, especially for women

FIGURE 4.15.
Testing the concept prototypes

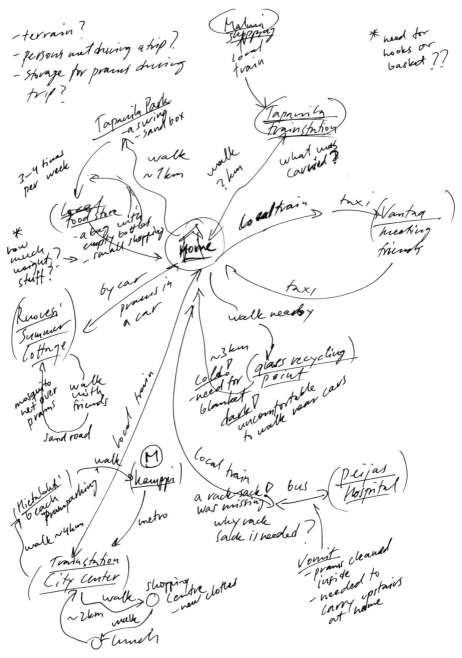

FIGURE 4.16.
Cognitive map of travelling with a child drawn on the basis of self-documentation.
The map was used as an aid in discussions with the family

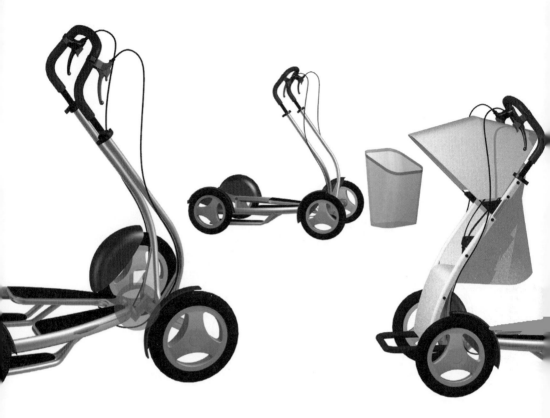

FIGURE 4.17.
Refined concept design with the new steering mechanism

following pregnancy. The similarity with other new types of exercise, such as Nordic walking, was mentioned. For families without a (second) car, the daily mobility range could be significantly extended with the new product. Children liked the low sitting position at the front of the kick-bike and the feeling of speed. The low centre of gravity was considered to be a positive safety factor. The carrying capacity was felt to be better than normal prams, such as when carrying shopping bags.

Early assumptions of the potential target groups were adjusted to cover a wider range of different uses. It seemed that the living environment was not a critical factor. The attitude towards physical exercise and outdoor

activity was more relevant to the interest in a product. Consequently, the concept was developed to meet more urban needs, such as fitting in a metro train, a bus, a local train and in shops. A previous design feature that allowed the kick-bike to be folded to fit in a car was discarded.

It became obvious that the new concept could not replace existing prams and could be a significant purchase for a young family. Therefore more emphasis was put on extending the life cycle of the product, including use before the child can sit upright and after the child is no longer being transported. Although the product image itself was regarded as appropriate, the design still seemed to have features that resembled those of aids.

The handlebars and pushing position had too much in common with the rollator walking aids for the elderly. However, the kick-bike was considered capable of competing with bicycles, because of the ease with which it could be controlled when riding with a child.

A major concept redesign was implemented after a new mechanical solution for the steering system was found. A patented structure solved many of the problems that had been identified, and it made the whole construction simpler, lighter and cheaper. There was no longer a need for a separate structure for the steering handles and seat. The overall handling was now improved and the image of the product was changed dramatically, resulting in a sportier and better balanced design.

The case study shows that a user-centred design can be adapted for use even in compact concept projects. By selecting the appropriate study method and integrating the end-user sessions, it took only 2 working weeks to complete the user-data gathering, interpretation and consolidation processes used to provide information for the design. In this case the framework for a concept had already been established when the project started. Through user involvement, it was possible to identify the relevant and crucial product features of this new product type.

4.7   References

1   **Gould, J. D., Lewis, C.** (1985): Designing for Usability: Key Principles and What Designers Think. Communications of the ACM, Vol. 28, No 3, March 1985.

2   **Beyer, H., Holtzblatt, K.** (1998) Contextual Design: Defining Customer-Centered Systems. InconMorgan Kaufmann Publishers, Inc., San Francisco.

3   **Sanders, E. B.-N.** (2001): Virtuosos of the Experience Domain. Proceedings of the 2001 IDSA Education Conference.

4   **Pine, J.B., Gilmore, J, H.** (1999): The Experience Economy: Work is Theatre & Every Business a Stage. Harvard Business School Press, Boston.

5   **Nielsen, J.** (1993): Usability Engineering. Academic Press, Chestnutt Hill.

6   **Schifferstein, H, N, J., Hekkert, P.** (2002): Designing consumer-product attachment. Proceedings of the 3rd Conference of Design and Emotion.

7   **Demirbilek, O., Sener, B.** (2001): A Design Language for Products: Designing for Happiness. Proceedings of Affective Human Factors Design 2001

8   **Nagamachi, M.** (1995): The Story of Kansei Engineering. Kaibundo Publishing, Tokyo.

9   **Vihma, S.** (1995): Products as Representations. University of Art and Design, Jyväskylä.

10   **Jordan, P. W.** (1999): Designing Pleasurable Products: An Introduction to the New Human Factors. Taylor & Francis, London.

11   **Papanek, V.** (1985): Design for the real world: human ecology and social change. Thames and Hudson, London.

12   **Klein, N.** (2000): No logo, no space, no choice, no jobs, taking aim at the brand bullies. Flamingo, London.

13   **Nieminen-Sundell, R., Panzar, M.** (2003): Towards an Ecology of Goods: Symbiosis and Competition between Material Household Commodities, in Empathic Design, eds. Koskinen, I., Battarbee, K., Mattelmäki, T. IT-Press.

14   **Kankainen, A.** (2002): Thinking Model and Tools for Understanding User Experience Related to Information Appliance Product Concepts. Dissertation, Helsinki University of Technology. Polytechnica Kustannus Oy.

15   **Kelley, T.** (2001): The Art of Innovation: Lessons in Creativity from IDEO, America's Leading Design Firm. Doubleday, New York.

16   **Mattelmäki, T.** (2003): Probes: Studying experiences for design empathy, in Empathic Design, eds. Koskinen, I., Battarbee, K., Mattelmäki, T. IT-Press.

17   **Jääskö, V., Mattelmäki, T.** (2003) Observing and Probing. Proceedings of Dppi 03: Designing pleasurable products and interfaces.

18   **Center for quality of management** (1995): The language processing method (LP): A tool for organizing qualitative data and creating insight. Cambridge.

19   **Buchenau, M., Fulton Suri, J.** (2000): Experience Prototyping. Proceedings of DIS 2000 Conference (ACM).

20   **Dreyfuss, H.** (2003). Designing for people. Allworth Press.

21   **Ehn, P., Badham, R.** (2002): Participatory design and the collective designer. Proceedings of the PDC 2002 Conference. Computer Professionals for Social Responsibility, Palo Alto.

22   **Gaver, W., Dunne, T., Pacenti, E.** (1999): Cultural probes. Interactions. Vol VI, No. 1 January/ February 1999, pp. 21-29.

23   **Keinonen, T** (2000): 1/4 Samara, in Designing Usability, ed. Keinonen, T. University of Art and Design publication B61, Helsinki.

24   **Keinonen, T.** (2003): 1/4 Samara – Hardware prototyping a driving simulator environment, in Mobile Usability: How Nokia changed the face of the mobile phone, eds. Lindhom, C. Keinonen, T., Kiljander, H. McGraw-Hill, New York.

25   **Lamble D., Laakso M., Summala H.** (1999): Detection thresholds in car-following situations and peripheral vision: Implications for the positioning of visually demanding in-car displays. Ergonomics 42 (1999): 6, pp. 807-815.

26   **Zwahlen, H. T., Adams, C. C., DeBald, D. P.** (1988): Safety Aspects of CRT Touch Panel Controls on Automobiles. Second International Conference on Vision in Vehicles, eds. Gale, A. G., Freeman, M. H., Haslegrave, C. M., Smith, P., Taylor, S. P., pp. 335-344.

27   **Kiljander, H., Järnström, J.** (2003): User Interface Styles, in Mobile Usability: How Nokia changed the face of the mobile phone, eds. Lindhom, C. Keinonen, T., Kiljander, H., McGraw-Hill, New York.

28   **Helle, S., Järnström, J., Koskinen, T.** (2003): Takeout Menus: Elements of Nokia's Mobile User Interface, in Mobile Usability: How Nokia changed the face of the mobile phone, eds. Lindhom, C. Keinonen, T., Kiljander, H., McGraw-Hill, New York.

5

Strategic Concepts in the Automotive Industry
– Volvo Case Study

Toni-Matti Karjalainen

5

Strategic Concepts in the Automotive Industry

— Volvo Case Study

Toni-Matti Karjalainen

5.1   Concepting in the car industry

The use of concept design for strategic purposes differs between industries and companies. Chapter 6 looks at the practices of concept design in a dynamic, unstable industry. This chapter provides an example from a more established market and product development environment – the car industry. As we will suggest, the activities characterised in Chapter 1 as "concept design for expectation management" and "concept design for innovation and shared vision" are of particular relevance in the automotive sector.

The basic product concept of a car has long since reached maturity. On a general level the technical and qualitative differences between competing products have been reduced. As a result, competition has moved firmly in the direction of symbolic meanings, created through branding, marketing communications and, more characteristically, through product design. Design has a special position in building deliberate references into products to reinforce intentional messages and, from the perspective of strategic brand management, to enhance recognition and differentiation. Creating distinctive and recognisable product messages through symbolic product design is a strategic cornerstone for an increasing number of manufacturers.

In addition to creating and maintaining a sense of distinctiveness, constant updates to product design are needed to ensure that companies remain competitive. These modifications can range from minor facelifts to the creation of new products or even to the complete reinvention of the overall design approach of the brand. Recently, many major car brands, such as BMW, Mercedes, Renault, Citroën and Nissan, have undergone a radical refocus of their design approach. The update to the Volvo design that forms the basis for the case study in this chapter is particularly interesting. More than ever before, manufacturers are striving to create more distinctive and more recognisable product identities. Brand-specific design seems to be a guiding approach in the industry, and concept design plays a significant role in this context.

Refocusing the direction of design and making it more distinctive involves a considerable challenge for car manufacturers. Within this established industry the nature of NPD is vastly different from that of many other consumer goods, such as mobile phones (as described in Chapter 6). Long development cycles and huge investments in NPD tend to make manufacturers wary of taking major risks. On the other hand, the increasing number of holistic platforms and component sharing practices that have spread throughout the industry have brought considerable reductions in research and development costs and, consequently, enabled the more effective development and production of larger product families consisting of a variety of different products. Despite the more flexible reality of NPD, the threshold for innovations and major design changes has remained relatively high. It is not in the interests of the majority of companies to alienate their often surprisingly conservative customer base with solutions that differ radically from their customary products. Although daring solutions may be found in concept studies, this risk-averse attitude is reflected in the more neutral designs of production models. The further the design process proceeds, the more the elements that have ventured too far from the mainstream become tempered. Designers are forced to compromise, partly because of technical and production requirements and partly as a result of a simple lack of courage on the part of the management.

This risk-averse reality has contributed to the birth of a strong tradition of presenting concept studies in public. This is a particularly characteristic aspect of the car industry in terms of its systematic and widespread nature. Presentations of concept studies play a central, eye-catching role at the major car shows such as Frankfurt, Paris, Geneva, Detroit and Tokyo. The life-size creations, which are often fully functional, may even attract attention away from production models. In concept cars, the emphasis is on creativity and the presentation of a bold, futuristic vision. The major car shows are important events for car companies where they can present their strategic identity. The public presentation of concept studies creates a specific arena for more creative testing and innovation. This chapter focuses on this type of design concept.

5.2   Strategic functions of concept cars

Concept studies have a major strategic importance for car companies. It has almost become a necessity for all the main manufacturers to present concept cars at the major car shows. There are, however, considerable differences in the approaches of different companies. For example, the strategy and, consequently, the image of French manufacturers have in recent years been influenced by their daring design studies. Renault is an example of a brand that, since 1990, has built a strong design identity through anomalous concept studies, and has later carried this through into its production models. Volkswagen, on the other hand, has traditionally put less emphasis on futuristic creations and focused on more realistic concepts that reflect the strategic identity of the brand. The approaches of both Renault and Volkswagen seem to be firmly based on their strategic choices. The car shows have also hosted a variety of controversial concepts that may present futuristic visions and interesting ideas, but are out of line with the strategic objectives of the company. In the worst case, strategically invalid concepts can create a feeling of inconsistency, and may even be harmful to the company's brand image in the long term.

The strategic potential of design concepts is the specific topic of this chapter. We suggest that, in addition to being an imperative for companies, public concepts can provide companies with a powerful strategic tool for promoting a brand identity and for product portfolio management. Concept studies that are independent of the "real" product development process offer possibilities for exploring expectations, shaping attitudes and paving the way for innovations, as well as maintaining the overall brand relevance. These three strategic roles of concept studies are briefly discussed below.

5.2.1   Exploring market reactions and expectations

Concept studies can be used to explore customers' and other stakeholders' expectations concerning a new model that is under consideration or already under development. Depending on the nature of the concept model, different types of feedback can be obtained. A concept can be designed that represents the initial design ideas of the forthcoming production model in a realistic manner. In this type of case, market response can be utilised to validate initial solutions or to identify elements that need to be redesigned. At the other end of the spectrum, companies may create "unrealistic" concept studies that are totally detached from the development of specific production models, in order to test reactions to different design solutions. Feedback from public concepts can therefore provide general indications of the direction of public reactions to the central characteristics of forthcoming production models. Nonetheless, when exploring reactions and expectations, it is important to bear in mind the limitations of this approach. It can be difficult to obtain general approval for innovative designs and new solutions when they differ radically from the contemporary trends. The public evaluates concepts by comparing them with existing products that are currently on the market. It can be difficult to obtain relevant feedback on a product with a design which is projected into a context that is years ahead. This is a major challenge, particularly for car manufacturers.

Market cycles and NPD processes cover long periods of time. The time span from the point when the initial idea for the new product is developed in the company to the point of market launch can be several years. Moreover, the design of a new product should maintain its relevance in the market for between 5 and 10 years. It is impossible for customers to anticipate world and market trends that are 5 to 10 years away and to relate the concept model to these trends. In such an established industry, customers have a strong reference base consisting of car models that are currently or have been on the market. Customers are more likely to notice, and oppose, design changes when they are radical in nature. Car manufacturers have recognised this problem, and process the feedback from design concepts, as well as the information obtained through customer interviews, accordingly.

5.2.2   Shaping attitudes and paving the way for innovations

Concepts not only have to anticipate possible future states; they can also be used to create future directions. The second main function of concept studies is the shaping of public attitudes, to allow the reactions to forthcoming production models and/or changes in the brand's design approach to be anticipated more easily. For instance, if a totally new car model is planned, or if the company is planning a radical reshape of its design approach, sending out hints of the expected changes can be useful. By doing this, the company can help to ensure that the eventual production model will not be too big a surprise for the market. Concept models do not even need to illustrate the forthcoming changes explicitly. Even if the eventual production models incorporate totally different design features, new concepts can communicate the new, more innovative approach of the brand and to that extent lessen the shock. If the public is accustomed to expect innovative solutions from the brand, new models with new design approaches are more likely to be approved. Renault and Citroën, for example, have communicated through many bold concept studies that design innovation is at the heart of their strategy. The public expects them to produce controversial products, and conventional solutions may disappoint their customers.

Design concepts, as vehicles of expectation management, are particularly important when totally new product categories are being created. Ideas for new categories are constantly being presented at car shows. Many of the recently established segments, such as sport utility vehicles (SUVs) and multi-purpose vehicles (MPVs), were, prior to the launches of production models, presented using concept studies. If the recent concepts are assessed and regarded as indications of future development, new variations on the existing categories and totally new car concepts will be realized in the years to come. Publicly presented concepts feed the dynamics of the car industry by creating design trends as competing manufacturers adopt some of the design features and elements and transfer them, consciously or subconsciously, to their own concepts and production models.

In addition to design and functional innovations, manufacturers also present various technological developments and innovations at car shows. It has become a common policy to present design and technology concepts in one package. Sometimes, however, pure technology concepts are on show, in some cases even based on an existing product.

5.2.3 Maintaining brand relevance

Design concepts can have a major symbolic impact on a brand's image. Not only do they provide designers with possibilities for testing the limits of creativity and conventionality, but they also function as a medium for communication, signalling that innovative thinking takes place within the company. Concept studies prove that a company is actively considering the future, which contributes to the maintenance of its brand relevance.

The pure publicity value of concept studies is immense. Concepts attract a great deal of attention at car shows, are reported on by the press and can also be used in marketing communications. Besides increasing the overall relevance of the brand in the public eye, concepts also reinforce customers' brand loyalty by raising their expectations of forthcoming production models. As a publicly presented concept can be understood as a direct reference to a future production model, customers may even postpone their purchase decision and wait for the new product. Even though the concepts do not necessarily make a direct reference to future production models, the symbolic value of the brand may be strengthened, and new associations are incorporated into every product in the brand's product portfolio.

5.3   Design concepts at Volvo

Concept cars are an excellent way of providing a glimpse of the future without being constrained by a specific design. They help us to make wise decisions in our development work ... By tradition, Volvo was an engineering-driven company, and concept cars were previously primarily regarded as a way of presenting new technology. However, as the automotive world and the media that cover it are visually focused, a concept car also needs an innovative design if it is to attract the right attention.

Peter Horbury, former design director of Volvo
(Volvo press release, November 11, 2001)

Concept design for expectation management has been an obvious part of the various public concept studies recently that have been presented by a variety of companies. As an example, we here take a closer look at Volvo's design concepts. The company's design approach underwent a radical redefinition in the 1990s[1]. The entire change was, in effect, initiated by a concept study, the environmental concept car (ECC). During this process, and in the intervening period, Volvo has increasingly used public design concepts for strategic purposes, particularly as vehicles of brand communication.

FIGURE 5.1.
The design of the Volvo Concept Car (above),
presented in 1980, was used later in several
production models, such as the Volvo 740 (right)

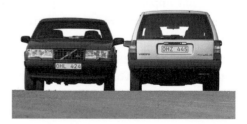

Prior to the "Revolvolution", as the strategic refocus has been called, Volvo did not consistently present new design ideas through public concepts. As the quotation above indicates, concepts were previously regarded as a basis for technological innovations. Design in general was not regarded as a strong competitive factor with respect to brand communication. In addition to the engineering-driven culture of the company, investments in design concepts were restricted by the small size and, consequently, the limited resources of the company.

The Revolvolution changed the situation. Product design became a key element of brand communication and the potential of design concepts was reconsidered. Consequently, the company has presented an increasing number of different design concepts at recent car shows. This development has also been strongly influenced by the increased R&D resources available to Volvo since 1999, when the Ford Motor Company acquired the firm. As a result of the Ford merger, the strategic significance of the concepts has increased. The Ford Motor Company currently has eight distinctive brands in its portfolio, great emphasis needs to be put on clearly differentiating them. Design concepts play an important role in creating and nurturing distinctive brand identities within the group.

However, a few interesting concepts can be chosen from Volvo's history to illustrate their role as vehicles of expectation management. Even though many historical Volvo concepts have been primarily regarded as technological concepts, they have also incorporated clear references to later production models. For example, the Volvo concept car presented in 1980 was designed to anticipate the general direction of Volvo design in the 1980s. The concept was used to prepare the public for a more cut-down and functional approach. The Volvo concept car led to the development of the Volvo 760, which was the first model in the new design era that characterised Volvo design for over a decade and was followed by many other models with a similar design approach (see Figure 5.1).

Twelve years later, a similar and even more significant change was initiated by the ECC that took Volvo into a new era. The ECC exhibited in 1992 at the Paris Motor Show, was seen as a strategic concept that represented the future design direction of Volvo. The Revolvolution, initiated by the ECC, was eventually revealed 6 years later when the first production model, the Volvo S80, was launched. In addition to the ECC, interesting concepts such as the environmental concept bus and the environmental concept truck were soon presented, not only to showcase new environmentally friendly technologies, but also to reshape the design of Volvo buses and trucks (see Figure 5.2). Of the more recent concepts, the adventure concept car was presented at the 2001 Detroit Motor Show and was a direct indication of a new Volvo model in the suv category, a predecessor of the XC90 production model launched in 2002.

Despite the increased number of concept studies being produced, they were not presented at every car show. Volvo's strategy when the company

FIGURE 5.2.

The ECC gave hints of the radical refocus of Volvo design when it was presented in 1992

had a new production model on display was to leave the concept studies at home – a relatively small brand cannot afford to risk damaging publicity for the production model.

In the car industry there seems to be a major need for ideation possibilities that are detached from the real routines of product development. Concept studies also play a similar specific role within NPD at Volvo. They function as a support for expectation management, as outlined at the beginning of this chapter. Concept studies are used to explore market expectations, to shape attitudes and to nurture brand relevance.

The first concrete stage in the Volvo NPD process is the business concept plan that describes a new product from a business perspective. In particular, it provides a picture of the intended customer group and its expectations concerning Volvo. This description, produced by a design team, business concept group and strategic planner, does not result in a concrete design concept at this stage; it is merely an imaginary description of the new product's character. A phase referred to as the "incubation period" follows the initial plan, which can take from 2 to as many as 6 years. The purpose of the incubation period is to allow the initial ideas to mature and to be refined into concrete product definitions. The most promising ideas lead to physical design concepts. Later, if the development of a production model is considered feasible, an actual product project is started. However, the incubation period does not delay the realisation of a new car, since a product can become reality at an earlier stage, depending on the development of the market and the structure of Volvo's long-term product road map.

The usefulness of innovative and experimental concept studies is highlighted by the incubation period. Concepts are an excellent way of making preliminary ideas more concrete and exploring various design solutions. They may never get as far as becoming an actual product project, but many of the ideas created for concepts are often used in other projects. The phase is a learning period for the people who take part in the product development process. Volvo designers consider that everyone should go through the incubation period to understand what the specific product is fundamentally about – this lays a solid foundation for the actual product project.

Concept design during the incubation period characteristically aims at a shared vision. The incubation period is, by definition, about providing time and space for ideas and thoughts to incubate at the back of the designers' minds. During this period, the designers are occupied with other development projects. Incubating concepts are rather different from the concepts created during the product project. Relatively loose parameters are specified for the design and creative solutions, with creativity and the radical testing of attitudes being regarded as virtues in the context of these studies. When the product project starts, many new requirements are identified that force the designers to compromise on a large scale.

In addition to concept design for expectation management, Volvo aims to create innovations using concept cars. These concepts are not created on the basis of specific business concepts with a specific production model in mind. A summary of Volvo's concept design activities is presented in Figure 5.3.

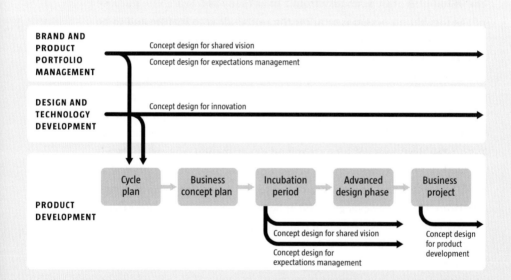

FIGURE 5.3.

The roles of concept design in Volvo NPD. The figure shows the links between business objectives and the goals of concept design at Volvo. Publicly presented concept studies typically result from the advanced design phase, and are primarily used as vehicles of expectation management

5.4   Revolvolution

We will now take a closer look at a specific Volvo design study, the ECC, which played an important role when Volvo was redirecting its design approach at the beginning of the 1990s.

Volvo attracted a great deal of attention in the car industry in 1998 when the Volvo S80 was launched. The car took a radical step away from the "boxy" design that had become characteristic of the brand in the 1980s and early 1990s. Through the Volvo S80 and its subsequent models, Volvo transformed its design identity from a world of simplified functionality and rationality into a domain of more powerful emotions. The new design approach was well received by the public. This was reflected not only in positive customer feedback and increased sales, but also in the large number of design prizes that the Volvo S80 and its successors were awarded. Most importantly, the new approach, although radically different from Volvo's previous design era, represented genuine continuity in Volvo's design history. Volvo's key identity attributes – safety and Scandinavian heritage – were taken as a starting point for the new approach. Direct and indirect references to historical Volvo models were also incorporated into the new models. Since the launch of the Volvo S80, consistent design cues have been utilised across the entire Volvo product portfolio. The current Volvo product family is one of the most consistent ranges within the automotive industry (see Figure 5.4).

All the characteristic design cues of the new Volvo design were first introduced at the 1992 Paris Motor Show by the ECC. The ECC was initially considered a technological concept, intended to present new environmental and safety solutions. The safety and environment groups at Volvo wanted to build the concept on the basis of the Volvo 940 estate model. This idea was understandably criticised by the design department, since a new concept would provide an excellent opportunity to indicate the future design of the Volvo brand. It was also realised that the Volvo design needed to be renewed. Eventually a decision was made to use this opportunity to build a vision of the future for Volvo design. The purpose was to convey to the public and the industry that Volvo was about to redesign its identity.

The ECC was planned as a quick design project that was assigned to Volvo's Californian design studio. The brief was simple: to envisage what Volvo would look like in the future. After the sketching phase, a review of alternative concepts was organised in the studio. The attention of Horbury,

S80

V70

S60

XC90

S40

V50

FIGURE 5.4.

The current Volvo product portfolio includes the S80, V70, S60, and XC 90, which were designed as a consistent family and were produced almost simultaneously by the Volvo design studio. The cars also share the same physical platform. The newest members of the Volvo family, the S40 and V50, have the same typical Volvo design cues as their "big sisters": a low-set, solid nose with a characteristic Volvo grille, powerful shoulders, a clearly visible V-shaped bonnet, sculptural tail lights and characteristic dynamic lines

Volvo's design director, immediately focused on the sketches drawn by an American designer, Doug Frasher. Horbury stated that he was struck by an immediate feeling that this was the road Volvo needed to take. In particular, he was attracted by the concept's strong design elements, which were characteristically modern but, at the same time, conveyed subtle references to certain important models from Volvo's history, such as the Volvo PV 444/544 and the P120 "Amazon". This was the type of association that Volvo was looking for.

As soon as the ECC was created, its design language was regarded as the direction for future Volvo design. Horbury wanted to transform the design solutions into a production model. Designers started work on a product that was planned to be built on a shared platform with a new Renault model. The planned fusion with Renault, however, unexpectedly fell through at the end of 1993. This caused a panic in the Volvo product development, since suddenly Volvo was lacking a badly needed competitive platform for a new product. The Volvo 960 was already becoming obsolete and the future of the company could not rest solely on the shoulders of the new Volvo 850.

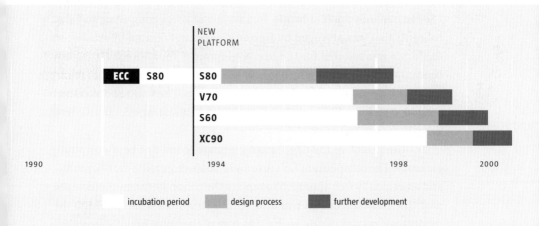

FIGURE 5.5.
The development periods for Volvo product families

In addition to the technological setback, the company was also lacking a proper business concept.

During the 1993 Christmas period, Volvo engineers developed the mechanical basis of the Volvo 850. A critical challenge was to develop a platform that would be suitable for several new Volvo models that were included in the cycle plan. Initial expectations were directed towards the first model, a large saloon, and another car in a new size segment, between the large (e.g. the Volvo 850) and the small (the Volvo 40 series) Volvo models. These products became the Volvo S80 and S60. The plan also included a new estate model, the Volvo V70, and its off-road XC70 version. In addition, a totally new Volvo model, the XC90, was later planned to exploit the booming SUV market.

The new platform was developed in record time, and the Volvo S80 project was finally properly launched at the beginning of 1994. Together with the Volvo S80, Volvo designers were also conceptualising its sister models. As shown in Figure 5.5, the whole product family was designed simultaneously in the Volvo design studio.

The starting point for the design of the Volvo S80 was to apply the design language of the ECC and, as an inherent attribute of the Volvo brand, to emphasise safety. During the conceptualisation period, several concepts were developed in all three of Volvo's design studios at the time: in Sweden, The Netherlands and California. In a review of three competing full-size concepts that was arranged by Horbury, a concept created by Frasher in California was selected for further development. The concept was a clear continuation of the design theme of the ECC. The design was frozen in the autumn of 1996, after a 2-year design process. The Volvo S80 was then presented to the public in 1998, followed by the annual introductions of the Volvo V70, S60, and XC90.

The launch of the new design language seemed to be a wise strategic move. The competition for customers' souls required a more dynamic and emotionally rich approach to replace the conservative methods of the past. The radical redesign was also skilfully implemented. Although the modernisation was clearly visible, the core of the Volvo brand was not forgotten. References to Volvo's strong heritage and a clear recognition of the Volvo brand were considered a self-evident starting point for the design. As Horbury concluded, the designers had to be very careful to maintain Volvo recognition, because it stood for many things that were valuable for the brand, even if it did not represent beautiful design. Volvo design cues seem to be highly functional from a strategic point of view. They are powerful and easily recognisable, but can also be adjusted to meet the needs of different model segments. In particular, the strategic significance of these elements is created by their references to the strong and credible heritage of the Volvo brand and to its core message: safety. Just as important as the heritage was the fact that the new products had a clearly contemporary appearance. According to Horbury the purpose of the references to the past was not to repeat, but to remind – the new Volvo design was not about retrospective design in the negative sense of the word.

5.5   Recent Volvo concepts – design in evolution

Recently, Volvo has introduced several new studies that anticipate the future direction of Volvo design: the scc (Safety Concept Car), the vcc (Versatility Concept Car), the ycc (Your Concept Car) and the 3cc (3 Concept Car).

5.5.1   Safety concept car – dynamic safety

The safety concept car (scc), which was introduced at the Detroit Motor Show in 2001, is an interesting example of the way in which Volvo's design language can be extended. It continued to explore the link between safety and excitement. The basic proportions of the scc were determined several months before the actual design work began by Frasher, Horbury and the engineering team. Four designers worked for a couple of weeks on their own proposals. In the review organised by Horbury, Stefan Jansson's proposal was selected for further development. Jansson, who was appointed as the designer responsible for the exterior of the scc project, was given a great deal of freedom and opportunities for exploration. But once again, safety was used as one of the main arguments for the car. The objective of the scc was to give rise to positive safety connotations rather than the traditional negative and boring image of safety. In order to enhance active safety, the concept introduced various technical innovations, new functions and also designs. The main theme of the scc was "superior vision". For instance, new designs for the A- and B-pillars resulted in improved vision (see Figure 5.6). Furthermore, a bright orange colour was chosen to enhance the visibility of the car in traffic and to communicate its sporty, dynamic character.

FIGURE 5.6.
Safety concept car

Although the final design of the study differed from earlier production models, the link was clearly visible. It includes all Volvo's major design cues. The extent to which the design of the scc will be used in future production models has yet to be seen. In any case, the impact of the concept on the Volvo brand has been considerable in terms of image and publicity.

5.5.2 Versatility concept car – Scandinavian functionality

The versatility concept car (vcc) was first displayed at the Geneva Motor Show in 2003. It was designed to celebrate the 100th anniversary of the Ford Motor Company. Once again, the concept seemed to explore the extendability of Volvo's design language by providing a combination of recognisable Volvo design cues and new innovative ideas. The exterior and interior design, were produced in Gothenburg and Barcelona, respectively. The car was primarily designed on computer and built entirely in-house, with the design, engineering and tooling departments working together. The study was completed extremely quickly. For example, it took only a few days to complete the entire interior; usually it takes about 9 months to develop a concept car at Volvo.

Once again, the car includes references to certain Volvo models of the past. For example, the grille and the four-light formation are a deliberate reinterpretation of the Volvo 164 that was launched in 1968[2]. The car was regarded as a "natural Volvo" that "could not be anything else"[2]. It incorporated familiar elements, such as the shoulders, V-shaped bonnet, recognisable grille and clear rear end which conveyed both safety and dynamism (see Figure 5.7). The interior design is also clearly recognisable. The deliberate goal was to use a simplified Scandinavian design language, as Joakim Karske,

FIGURE 5.7.
Versatility concept car

the chief designer of the VCC interior, explained. The Barcelona studio aimed to combine Scandinavian purity with a Mediterranean twist.

5.5.3   Your concept car – feminine interpretation of Volvo design

Volvo's presentation of a concept study, the your concept car (YCC), aroused a great deal of interest at the 2004 Geneva Motor Show, and subsequently in the automotive design world. The punchline of the concept was related to its design and development – the car was designed almost entirely by women.[4,5]

The beginning of the YCC dates back to 2001, when a seminar was organised at Volvo to discuss how Volvo could appeal more effectively to women customers. A challenging idea emerged: what would a car look like if it was designed only by women? At the end of 2002, a team was formed to take the idea further. The project lasted only 15 months and was completed within a budget of 2.6 million euros, which made it the most inexpensive concept car in Volvo's history. According to a YCC brochure, the design team was given

FIGURE 5.8.
Your concept car

"a free hand to develop a concept car capable of winning the approval of that most demanding Volvo customer category of all – the independent female professional". Volvo's customer research had shown that female buyers in the premium segment, in addition to performance, prestige and style, also wanted clever storage, easy access and exit, good visibility, options for personalisation, minimal maintenance and easy parking. These attributes were the starting point for the YCC.

These requirements were clearly reflected in the main features of the final design. Good visibility, for instance, was promoted by solutions that help a driver to see the corners of the car and to have a good view over her shoulder. Personalisation options were incorporated into the interior, including exchangeable seat pads and carpets. Gull-wing doors facilitate easy access and entry. The result of the female input to the project was said to be most obvious in the interior and in the solutions for the use of space. In addition to being targeted at women, it was also essential that the YCC had a strong sense of "Volvoness". Designers followed the Volvo design principles, and the product became a true representative of Volvo design philosophy, with direct references to the Volvo P1800, one of Volvo's truly classic models. Its design incorporates muscular but elegant elements. The V-bonnet, shoulders, sculptural rear lights and recognisable Volvo nose are all in place (see Figure 5.8). Moreover, "Scandinavian lightness" was again noted as inspirating the choice of materials and surface finishing.

5.5.4   Three concept car – the new small Volvo?

The three concept car (3cc), the most recent Volvo concept study at the time of writing, was presented for the first time at the Michelin Challenge Bibendum in Shanghai in October 2004. It was the result of a project where designers, engineers and business people from the Volvo monitoring and concept center in California investigated new innovative solutions for an ecological car of the future. The task was to create a "future-proof concept" that would enhance sustainable mobility[3]. This referred to a car that would not only be fuel-efficient, versatile, comfortable and safe, but also exciting to drive and look at[3]. The emotional appeal to all the senses was seen as an important design attribute. The resulting product is similar in size to a classic two-seater sports car but incorporates a unique two-plus-one solution in terms of space and seating arrangements. The 3cc provides seating for two adults in the front and a rear-seat solution for an additional adult or two children. Although it is a small car, the Volvo 3CC has been designed to feel spacious as a result of its organic lines and light colours. Moreover, many design details, such as the low-profile A-pillar, transparent roof panels and the position of the seats create a sense of openness inside the car[6].

Many technical innovations were included to support the key theme of the concept: efficient and sustainable mobility. The emphasis was put on efficient aerodynamics for the bodywork, lightweight body materials were used and an electric power train was developed for the concept. Safety was naturally regarded once again as a core attribute. The challenge was substantial, because a high level of safety was now being incorporated into a small-segment vehicle. As a result, an intelligent concept for handling

FIGURE 5.9.
Three concept car

the incoming forces in a frontal collision was developed. As can be seen in Figure 5.9, the common Volvo design cues have once more been used in a new interpretation.

There has been speculation that the 3CC is a strong indication of the new small Volvo model that is currently being developed and will presumably be launched within a couple of years. The same predictions were made in connection with the YCC and SCC. If they are seen as vehicles of expectation management, this could very well be the case. The comments of Lex Kerssemakers, senior vice president of brand, product and business strategy at Volvo, also point in this direction[3]: "We think the Volvo 3CC opens a door into the future, and we will develop the concept further".

5.6   Concluding remarks

The public presentation of concept studies is widely used by car companies to explore market expectations and reactions, to shape public attitudes, to pave the way for innovations and to maintain brand relevance. The Volvo case study is an interesting example of the role of concepting in the automotive industry.

The variability of the Volvo design language is illustrated in recent concept cars, which incorporate familiar elements that refer to recent production models, but that have been interpreted differently. Concept studies for expectation management, in the case of new Volvo concepts, involve examining the modifiability and expandability of the design approach and visualising future versions of it. Concept cars function as platforms for discussion within the company[2]; they are concrete examples with physical shapes that form the basis for discussions about future directions.

5.7   References

1   **Karjalainen, Toni-Matti** (2004): *Semantic Transformation in Design. Communicating strategic brand identity through product design references.* University of Art and Design in Helsinki.
2   **Horrell, Paul** (2003): Volvo: The Next Generation. VCC Concept Car. *Car*, June 2003. 104-109.
3   http://www.conceptcar.co.uk/news/design/cardesignnews28.php
4   **Möller, Simon** (2004): Back to the Future. *Design Report*, 10/04. 62-65.
5   **Your Concept Car.** *Design Management News & Views*, Vol XVII, No 1. Design Management Institute. p.3.
6   **Volvo press releases**, March 1 2005 and January 10 2005.

6

Concepts in Uncertain Business Environments

Anssi Tuulenmäki

6

Concepts in Uncertain Business Environments

Anssi Tuulenmäki

6.1   Introduction

When a company is aiming to realise its conceptual ideas, the goal of the conceptualisation phase is to determine whether or not a concept should be developed into product. The information needed for well-founded decision-making includes, at the very least, the product's competitive advantage, its target customers, technological feasibility, market context, costs and risk assessment, as well as manufacturability and timing considerations[1–3]. Thus, product development concepting provides information for the implementation decision, and it is only after this decision has been made that the launch-oriented product development process can start. This is how decisions about new products that offer incremental improvements have traditionally been made.

Highly uncertain and/or rapidly changing business environments pose a fundamental challenge for the early phases of NPD. Companies need to decide how the NPD process – and the concept design as a key phase of it – should be organised in environments where accurate information is not available during the early phases of the process. We argue that in situations such as this, the roles of concept design and product development will become mixed. The products being launched can be seen and treated as concepts, since they are part of the process of creating information for bigger investment decisions. They are used as tools for improving the company's competence in operating in the new market and for influencing the customers' perception of the emerging product categories. Consequently, the products being launched are used to achieve several of the goals characteristically associated with concepts (see Section 1.7).

In order to address the topic, we look first at the influence of uncertainty on NPD processes. Second, we consider several different methods in order to illustrate how concepting can be carried out in dynamic and uncertain environments. Each method has its advantages, which depend on iteration costs, concept and product characteristics, and market dynamics. We also give a few real-life examples, showing how Nokia has reduced uncertainty in the mobile communications business.

6.2   Uncertainty and new product development

Many product design researchers have provided process models for NPD, technology development and commercialisation. The most widely used process models are variations on the Stage-Gate™ process[4]. The basic assumption with Stage-Gate™- or waterfall-type models, as they are also called, is that all the key issues can be defined, planned and scheduled during the concept design phase at the start of the product development process. In other words, this type of process model relies on predictable information and thus anticipated outcomes. However, anticipation is an effective working mode only in stable markets where customer preferences and technologies are well known and do not change fundamentally during the product development process. A proper concept design phase is based on information that has been carefully gathered and thoroughly analysed. The level of certainty does not increase substantially during the concept definition process, because it is already high at the beginning of the process. Also, the earlier the concept is frozen, the faster the downstream process can be implemented (see Figure 6.1).

In more uncertain, complex and dynamic business environments, accurate information about the key features, performance variables and customer preferences is not readily available during the very beginning of the process. This kind of uncertainty has direct implications for the NPD process. As a result of the lack of reliable information at an early stage, the

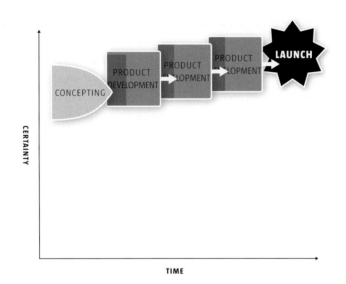

FIGURE 6.1.

Successive phases of product development in established markets. The level of certainty is already high in the first phase and does not increase significantly during the process

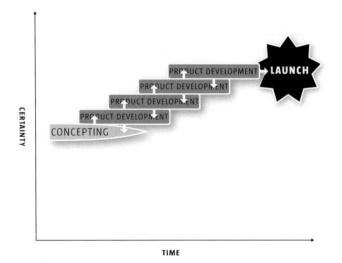

FIGURE 6.2.

A flexible NPD process in an uncertain market. The phases overlap and the level of certainty increases significantly during the process

key process factors for successful results are planned flexibility and reaction capabilities[5,6].

There are at least two challenges posed by more uncertain environments for Stage-Gate™ models. First, greater uncertainty challenges the assumption that all the necessary information about potential design choices is known or can be discovered, when the concept is defined at the start of the NPD process. In uncertain and dynamic environments, companies cannot predict the impact of potential design choices in advance. Therefore, there is a greater need to keep the product concept open to change. To achieve this, the early design stages must overlap those that follow (Figure 6.2). The later phases must start before the earlier ones can be completed, in order to allow acquired information to be used to create an iterative process of product definition. A decision made in an early phase must be tested in the next one so as to understand its impact. In the Stage-Gate™ approach, phases with low interdependence might be carried out concurrently due to speed imperatives. In more uncertain business environments phases are run in parallel, mainly because the earlier phases cannot be completed until the information from the following phases is available.

Complex, systemic products pose the second challenge for Stage-Gate™ models. Feedback on how a product performs as part of a system is not available until the later stages of a project, when the characteristics of each module and subsystem have been developed. Therefore, feedback must be obtained from early system-level tests, implying that the later stages must also overlap. Hence the system-level performance should be first tested before the modules are complete. Moreover, in the most flexible processes all the phases will overlap, thereby allowing feedback from system-level tests to have a direct impact on the evolution of the product concept. Again, in stark contrast to the Stage-Gate™ approach, the later one is able to freeze the concept the better, since uncertainty is reduced significantly during the process.

In the most challenging end of the uncertainty continuum is discontinuous or radical innovations in dynamic, rapidly changing business environments where technologies, markets and customer preferences are emerging, fluctuating and mostly unknown. Since product development is, as Eisenhardt and Tabrizi[7] express it, "a very uncertain path through foggy and shifting markets and technologies", traditional market testing is of limited use[8,9]. Similarly, being flexible during the initial NPD process is insufficient. Rather, something has to be launched on the marketplace before accurate information regarding customers and features can be obtained. Information cannot be gathered before it is created, and so, even the first launched product or products can be seen as concepts and thus as part of the concept design phase. The certainty level does not increase substantially until the first version is launched and real-life experiences have been analysed. New variants can then be developed and launched on the basis of this knowledge (see Figure 6.3).

Naturally, this last approach is useful only if management is confident enough to commit to comprehensive experimentation or extensive concepting, even though it does not know exactly which kinds of concepts will turn out to be the winning ones. As Lynn et al.[9] argue, "probing with immature versions of the product makes sense if it serves as a vehicle for learning." The key challenges in this kind of iterative process are the speed and costs of market and technology knowledge accumulation, rather than the first launch being exactly correct.

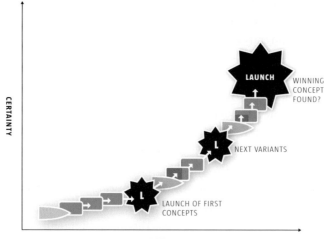

FIGURE 6.3.

In a very uncertain business environment, even the launched products can be seen as concepts because they serve as vehicles for learning

6.3   Product development concepting in uncertain conditions

As discussed above, sometimes even the most flexible NPD processes are insufficient for achieving optimum design and market results for the first launch. Market success might not be achieved with the first generation of products, but be only achieved after several launches. The goal of expeditionary marketing[10] is similar to the goal of concepting: to determine the precise direction in which to move. It defines the particular configuration of product characteristics that customers really appreciate, the winning business logic and the combination of price and performance that will open up a new competitive space. There are two ways to increase the number of hits in new business development[10]. One is to try to improve the odds on each individual bet, which is the hit rate. Individual personality-related issues have a powerful impact on the success of new business developments, especially during the very early stages when only a few people are involved in the process[11]. Selecting the right people for the right roles seems to improve the hit rate dramatically (see Chapter 2). The other way to increase the number of hits is according to Hamel and Prahalad[10] "to place many small bets in quick succession and hope that one will hit the jackpot". If the goal is to accumulate understanding as quickly as possible, a series of low-cost, fast-paced market incursions – expeditionary marketing – can bring the target into view more rapidly. When the target has finally been identified, it is time to move from concepting mode to business-as-usual product development. Then, a significant amount of money can be invested in each of the projects with the assumption that every project will bring a good return on the investment.

Leonard-Barton[8] divided experimentation strategies into three groups: Darwinian, product morphing and vicarious. In *Darwinian selection*, (Figure 6.4) a company launches multiple new product offerings simultaneously to see which has the greatest market appeal. Traditionally, Japanese companies have favoured this type of approach. For example, Sharp tried several different types of personal digital assistant (PDA) simultaneously in 1993 and 1994, and Sony developed hundreds of Walkman models to identify customer preferences[5,12].

In the *product morphing* method (Figure 6.5), initial product features evolve as consumers adopt and adapt the new products, and manufacturers implement the new technology offerings. General Electric used this approach in its computed axial topography (CAT) scanning business from

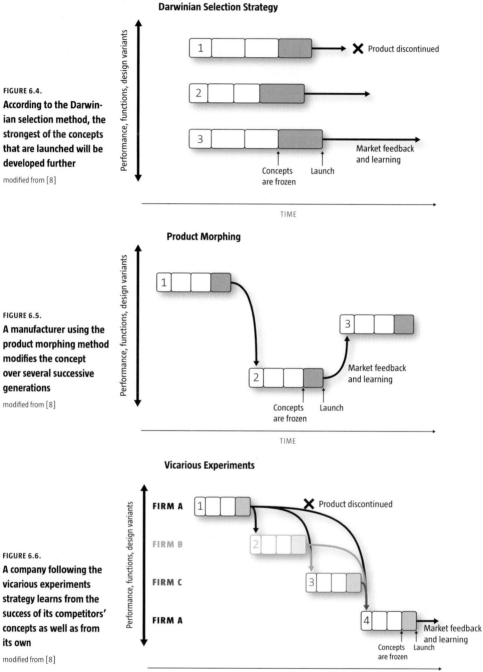

Darwinian Selection Strategy

Performance, functions, design variants

1 — Product discontinued

2

3 — Market feedback and learning

Concepts are frozen   Launch

TIME

FIGURE 6.4.
According to the Darwinian selection method, the strongest of the concepts that are launched will be developed further
modified from [8]

Product Morphing

Performance, functions, design variants

1

2 — Market feedback and learning

3

Concepts are frozen   Launch

TIME

FIGURE 6.5.
A manufacturer using the product morphing method modifies the concept over several successive generations
modified from [8]

Vicarious Experiments

Performance, functions, design variants

FIRM A   1 — Product discontinued

FIRM B   2

FIRM C   3

FIRM A   4 — Market feedback and learning

Concepts are frozen   Launch

TIME

FIGURE 6.6.
A company following the vicarious experiments strategy learns from the success of its competitors' concepts as well as from its own
modified from [8]

1975 to 1985, as did Motorola for the emerging hand-held mobile phone business from 1973 to 1984[9].

In the *vicarious experiments* method (Figure 6.6), companies learn from competitors and from their potential customers; the best elements are then incorporated into their own new offerings. This strategy includes learning from others' mistakes and is therefore based on a wait-and-see attitude. This is surely the cheapest kind of market experimentation, unless the pioneers succeed in creating dominant standards and/or dominant designs[13] or unless the market is strongly influenced by network externalities and/or high switching costs. In this case the result might be a technological or market lockout[14].

Figure 6.7 shows another possible strategy, termed the *variants strategy*. In this case experimentation – or, when used in more mature markets, segmentation – is based on variants of an original mother product. Instead of starting an NPD project from scratch, the new variants use parts of the first product, but new technologies, new designs and/or new features are introduced on top of it rapidly and cost-effectively. Thus, one could argue that experimentation continues after the mother product is launched. The best parts of the product from the customer's point of view, or the parts that have been expensive for the company to develop, are used in several variant launches. The launches aim at probing evolving customer preferences and to detect potential new concept ideas. Naturally, this approach calls for a high degree of modularity in product architectures.

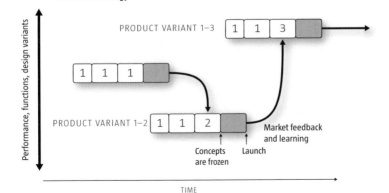

Variants Strategy

FIGURE 6.7.

The Variants strategy is an approach where the company generates several variations of a mother product so as to improve the learning process.

Figure 6.8 illustrates how Nokia used the variants strategy when targeting the mobile gaming market. The Nokia 5510 was basically a 3000-series phone rotated through 90 degrees, with a larger memory for games and MP3 tracks and with a keyboard divided onto both sides of the screen. The Nokia designers played down the phone-like features of the Nokia 5510 in order to make the gadget look more like an entertainment or a games station. By utilizing variants strategy, Nokia was able to test customers' reactions to the new positioning, design, features and sales channels. From a manufacturer's perspective, selling games and music is a completely different proposition from selling hardware, as was the case in the basic mobile phone segments. With the experience and knowledge gathered from the Nokia 5510, it was easier for Nokia to launch the N-Gage, which was clearly profiled as a gaming device. However, the N-Gage exhibited a few drawbacks that are critical in the gaming segment. For example, players could not change the game cartridges without removing the battery. Once again, the company learnt from its experiences and launched the new and improved N-Gage QD.

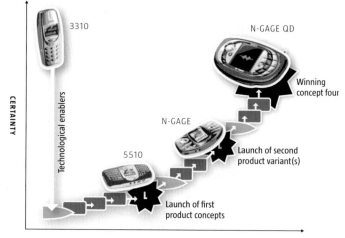

FIGURE 6.8.

Nokia learned about the mobile gaming market using the variants strategy

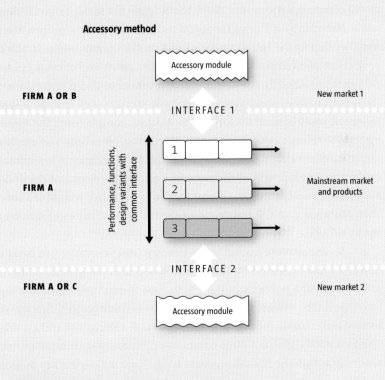

FIGURE 6.9.

Learning from market – the accessory method in which the mainstream product is linked to new markets by interfaces

Another experimenting strategy, which also calls for modularity, is here referred to as the *accessory method* (Figure 6.9). There, a company approaches very uncertain markets by organizing an interface that links the new offering to something that has already been widely adopted. A new concept can be linked and positioned as an accessory both by established players and by newcomers, both in intra- and interfirm settings.

Nokia's approach to the satellite communication business is an example of how established players can utilise an accessory strategy when approaching an uncertain and highly risky new business environment. In the end of the 1980s, satellites were considered one possible method of enabling global mobile communication. In 1988, Motorola engineers had conceptualised a network of low-earth-orbit satellites that could provide global communications service[16]. The satellite business, named Iridium, was launched in 1991. During the 1990s, Iridium and several similar ventures were developing the satellite network and satellite phone business. Market projections showed significant future business potential. For example, Merrill Lynch predicted (in April 1998) global revenues of US $31 billion in 2007 from the satellite-based mobile telephone services, and that 40 million terminals could be sold in 2007[17]. The technical complexity of the Iridium system, in terms of developing, manufacturing, launching and positioning 66 satellites circling the planet at 17 000 miles an hour in low-altitude orbits, the development of the network switching system and the design and development of the actual telephone was compared to that of a moon shot. Motorola CEO C.B. Galvin called the Iridium as "the eight wonder of the world"[18].

So, the satellite market in the late 1990s was characterised by astronomical initial investments for the networks, huge predictions of market growth and high publicity of the efforts. Due to enormous costs and high uncertainty, the market was also perceived as extremely risky. This was the picture Nokia had in 1997 when it decided to take a closer look at the market. The idea itself – global mobile communication enabled by satellite – was naturally within the Nokia's business focus. It sat well in the key message and promise of Nokia: connecting people. Even though Motorola launched the first models of satellite phones in 1998 and the Iridium service started at the end of that year, Nokia listened to the critical voices that were asking if those businesses would ever make enough profit to cover the billions of dollars invested. Nokia perceived the satellite business as too risky and thus decided not to start producing actual satellite phone terminals. Instead, the idea of an accessory module linked to the standard Nokia terminals using the Bluetooth protocol was developed and patented. The accessory device would communicate with the satellite network and then with any Nokia terminal over a Bluetooth link. A user could leave the satellite device outside (similar to a satellite television antenna, since a line-of-site was needed between the handset and orbiting satellites) and use the standard Nokia terminal inside. With these decisions, the core mainstream products of Nokia were left untouched. The open interface provided by Bluetooth was the only link to the mainstream products. To integrate an accessory, no large, complicated changes were needed to existing platforms, which certainly decreased the risks. It was seen as an add-on business to the Nokia mainstream terminals, that provided yet another feature.

By using the accessory strategy, Nokia got a cheap, first-row seat to see how the satellite communication market developed without risking its mainstream business. To reduce the risk further, even the development and manufacturing of those modules was assigned to external parties through the joint development agreements,with Nokia investing only one full-time person in the project. Soon after Nokia's project started, the satellite operators ran out of money and entered into ownership restructuring processes. Nokia immediately put the project on hold after spending totally only several man-years on the satellite business. In contrast, Motorola lost US $1–2 billion and investors lost even more[18,19] as the satellite communication business never lived up the huge promise.

The early PDA market is another example of how the accessory strategy can be utilized when entering new and uncertain markets. In this case, the accessory approach was used to increase the credibility of a new category by linking it via an interface to the well-established personal computer category. In 1992, John Sculley, a former chairman and chief executive officer (CEO) of Apple, coined the name "personal digital assistant" in a speech. Intense discussion started in the newspapers and the computer magazines on this new category that promised to combine features from the three most widely used information technologies: pen and paper, personal computer and telephone. Some large and prestigious companies, such as Apple, Tandy, Casio, IBM and Microsoft, made significant investments in the technology and committed themselves to producing hardware. Announcements and the first PDA product introductions in 1993 created the large amount of technology publicity round the category as a whole. Similar to the satellite phone business, studies forecasted huge sales for PDAs in the near future. However, after a very short period of initial publicity, it was clear that the first round of PDAs fell short of the expectations. Consequently, in 1994 the market showed signs of cooling down. Compaq Canada indefinitely postponed the release of its hand-held mobile communicator, and Microsoft admitted it would delay shipment of its WinPad device until the second quarter of 1995[20]. In early 1995, Motorola also announced that it was being forced to downsize due to slow PDA sales[21]. By 1996 the downsizing and discontinuation announcements by several companies challenged the creditability and viability of the entire category. With hindsight, the key problem with all the early PDAs was why anyone would pay US$600–1000 for a portable PDA or a handheld PC with limited functionality when you could get a portable laptop that offered so much more for only a slightly higher amount.

Palm shipped its first Palm Pilot machines in April 1996, which changed everything. Within 6 months, the pocket-size, pen-based PDA owned half of its market, and by November 1997 it had shipped one million units[22]. Palm termed its own product an "electronic organizer" and positioned it as a PC accessory. Users had to push just one button to synchronize the Palm Pilot with a PC. The focus was on managing and accessing information rather than creating and editing documents. The PC-accessory reasoning had certain advantages. First, Palm understood that there was no point making another portable gadget that was a "full extension of a desktop PC", as most of the other PDA makers labelled their devices. There was also no use trying to replace laptops or desktop computers. Such a PDA based on the affordable, available technology would be too large to carry easily and still too small to be really useful. Moreover, the price of such a device would be too close to that of a laptop computer. Rather, by positioning the new electronic organizer as a PC accessory, Palm utilized the well-established PC category to bring credibility to the new offering and thus reduced the customers' perceived risks of the new gadget.

Since then, PDA makers have been active in the accessory strategy by linking their PDAs via different interfaces to a large variety of external accessories: modems, digital cameras, MP3 music players, global positioning satellite navigators, etc. After manufacturers have seen which accessories become most widely accepted, they integrate them into the basic PDA product. This is what is happening currently with digital cameras.

6.4   Finding the optimum concepting strategy

How many new versions of concepts should a company introduce? The optimum number depends on the iteration costs, the product, the concept and the market characteristics, as well as on the competitive pressures. In markets with few rival products, the importance of monitoring and learning from the market using a greater variety of products outweighs the benefits of culling[15]. Similarly, product variety is more valuable when uncertainty makes accurate prediction difficult. Furthermore, the benefits of launching concepts decrease when the costs of iteration increase.

When competitive pressures are low and the cost of iteration is high, as was the case with GE's CAT scanning business[9], then the product morphing strategy might be the ideal solution. The company had a 10-year opportunity window to introduce new, improved but very expensive models and to find the winning concept.

If products and concepts are expensive to develop, but the market is much more dynamic, product morphing might be too slow approach. Instead, the variants strategy could be more appropriate, as companies can use the expensive and carefully developed features in several variant models and then simply add new technologies or features to these models. Of course, there is a delicate trade-off between the benefits and risks of this type of approach, because developing a product architecture that is modular and flexible enough to accommodate these add-on variants could be expensive and time-consuming.

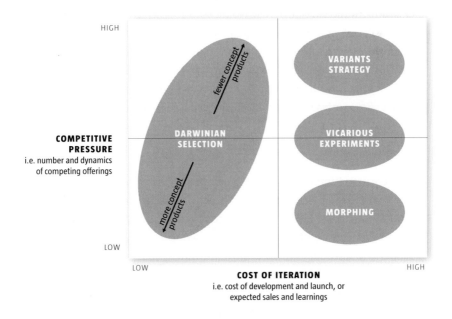

FIGURE 6.10.

The optimum conceptualisation approach can be chosen on the basis of competitive pressure and the costs of concept iterations

When the expected gains from new concepts are at their lowest, because the costs of iteration are high and yet the dynamics of the new market seem moderate, it makes sense to let others blaze the trail. In this case the vicarious experiments or accessory strategies might be the most appropriate ones, with a higher uncertainty and a higher cost making the accessory method the preferred strategy. In the vicarious experiments strategy, a company may introduce a product onto a new market, but the emphasis will be on market observation. In the accessory strategy, a company does not commit to produce anything – it just offers the interface.

If the concepts are relatively cheap to develop, as was the case with the Sony Walkman[12], then the Darwinian selection strategy might be appropriate. The gains from a large variety of concepts decrease when the amount of competing offerings or the cost of iteration increases. Figure 6.10 provides a summary of the concepting strategies.

In reality, companies often use a dynamic combination of the strategies described above. Again, Nokia provides a good example of this. When it was introduced in 2001, the Nokia 5510 looked like no previous Nokia phone, despite it inheriting most of its components and technical solutions from the standard Nokia 3300-series phone. The experience gained from the Nokia 5510 helped the company to develop three new phone categories: the Nokia 3300-series media phone, the Nokia N-Gage games phone and the Nokia 6800-series business phone (Figure 6.11).

6.5   Conclusion

Conceptualisation is a challenging task in a highly uncertain environment. The methods and strategies presented in this chapter are based on the assumption that the most effective and often the only way to obtain enough information for well-founded decision-making in highly uncertain environments is to introduce products into the evolving marketplace. For all of the strategies, the most important factor is not being right the first time, but the pace and cost at which the company recalibrates its product offering and learns from a new market. Faster learners might gain a competitive advantage in two general arenas. First, because experimentation gives them first-hand experience about customer preferences, value expectations and product characteristics, they are more likely to be in a better competitive position to develop higher volume products in the specific market domain.

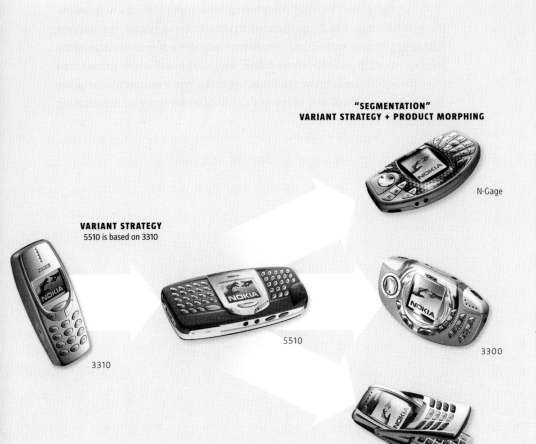

FIGURE 6.11.
Nokia developed new mobile terminal concepts using a combination of variant and product morphing strategies

In other words, they know exactly where to aim with their higher volume products. Second, companies active in experimentation may gain advantages in experimentation and concept creation itself. They may develop capabilities for faster learning and faster adaptation, which allows them to be more effective when faced with uncertainty and constant changes.

Design plays a key role in producing the winning combinations in new market situations. The design process, in the same way as any other process that aims to create something new, always includes a degree of uncertainty. Inquiry, iteration and experimentation, which are vital for new market creation, are also at the heart of the design process. When used effectively, the design process is thus one of the key capabilities for companies operating in uncertain environments.

6.6   References

1   **Kim, Jongbae & Wilemon, David** (2002); Strategic issues in managing innovation's fuzzy front end. European Journal of Innovation Management, Vol 5, No 1, pp 27-39.

2   **Ulrich, Karl T. & Eppinger, Steven D.** (2000); Product design and development, 2nd ed. Irwin-McGraw-Hill.

3   **Khurana, Anil & Rosenthal, Stephen R.** (1999); Towards "holistic" front ends in new product development. Journal of Product Innovation Management, 15:57-745.

4   **Cooper R.G.; Edgett S.J.; Kleinschmidt E.J.** (2002); Optimizing the stage-gate process: what best-practice companies do – I. Research Technology Management 45 (5), Sep-Oct, 21-27.

5   **MacCormack, Alan; Verganti, Roberto; Iansiti, Marco** (2001); Developing products on "Internet time": The anatomy of a flexible development process. Management Science, Vol 47, No 1 January, pp. 133-150.

6   **Verganti, Roberto** (1999); Planned flexibility: Linking anticipation and reaction in product development projects. The Journal of Product Innovation Management, Vol 16, pp. 363-376.

7   **Eisenhardt K.M. & Tabrizi B.** (1995); Accelerating adaptive processes: product innovation in the global computer industry. Administrative Science Quarterly, Vol 40, pp 84-110.

8   **Leonard-Barton, Dorothy** (1995); Wellsprings of knowledge. Harvard Business School Press; Boston.

9   **Lynn G.S.; Morone J.G.; Paulson A.S.** (1996); Marketing and discontinuous innovation: The probe and learn process. California Management Review, 38, No 3: 8-34.

10 **Hamel, Gary & Prahalad C.K.** (1991); Corporate imagination and expeditionary marketing. Harvard Business Review, July-August, pp. 81-92.

11 **Stevens, Greg A. & Burley, James** (2003); Piloting the rocket of radical innovation. Research Technology Management 46, No 2: 16-25.

13 **Anderson P. & Tushman M.L.** (1990); Technological discontinuities and dominant designs: a cyclical model of technological change. Administrative Science Quarterly 35, 604-633.

12 **Sanderson, Susan & Uzumeri, Mustafa** (1995); Managing product families: The case of Sony Walkman. Research Policy, Vol 24, pp.761-782.

14 **Schilling, Melissa A.** (1998); Technological lockout: An integrative model of the economic and strategic factors driving technology success and failure. The Academy of Management Review, Vol 23, No. 2, pp 267-284.

15 **Sorenson, Olav** (2000); Letting the market work for you: An evolutionary perspective on product strategy. Strategic Management Journal, Vol 21, pp.577-592

16 **Higgins, Kevin T** (1999); The anytime, anywhere phone company. Marketing Management, Vol 8, No 1 Spring; Chicago.

17 **Middleton, Bruce S** (1999); Reality check looms for GMPCS. Satellite Communications, Vol 23, No 1 Jan; Atlanta.

18 **Laing, Jonathan R** (1999); Lost in space. Barron's, Vol 79, No 31 August; Chicopee.

19 **Crockett R.O & Yang C** (1999); Why Motorola should hang up on Iridium. Business Week, Aug 30; NY.

20 **Hwang, Diana** (1994); Suppliers shift focus to boost PDA sales. Computer Reseller News, September 26; Manhasset.

21 **Blodgett, Mindy** (1995); Slow PDA sales force Motorola to downsize. Computerworld, August 21; Framingham.

22 **Graves, Lucas** (1998); Edward Colligan: VP of marketing, 3Com's Palm Computing. MC Technology Marketing Intelligence, February; New York.

7 Vision Concepts

Mikko Sääskilahti, Roope Takala

7   Vision Concepts[1]

Mikko Sääskilahti, Roope Takala

7.1   Concept development outside the research and development time frame

Almost all companies find themselves at some point in a situation where they are unable to provide the right products to address the existing market demand[2]. The company has perhaps failed to anticipate market changes, and so when it eventually comes face-to-face with the new competition it has no time, no skills and no resources available to respond to it.

Vision-based concept creation and concept development outside the normal R & D time frame is a means of preparing for changes in the business environment. Vision concepting (see the description in Chapter 1) aims to sketch out future products and product portfolios far beyond the normal product development perspective, up to 15 to 20 years ahead. These concepts can be used to answer the question: "What kind of business will we be in in 15 years time?" Creating a vision concept gives a view beyond the blinding, close-range blur, which includes factors such as the current competition,

---

1   This chapter is based on an original text[1] published in Finnish. We hereby acknowledge the authors of the original text: Ville Kokkonen, Markku Kuuva, Sami Leppimäki, Ville Lähteinen, Tarja Meristö and Sampsa Piira.

the prevailing operational practices and the existing set of resources and competences. It enables the taboos of today's business to be challenged and modified in order to identify possible future paths.

This book discusses several reasons for generating concepts. One of these is that, by introducing concepts such as Volvo's concept cars (see Chapter 5), the manufacturer can propose new ideas that the public can become familiar with and take on board. The company can also study the reactions and feedback from the market. Concepting can be used to generate a dialogue between the company and its customers. In the case of vision concepts, the customer base for the futuristic ideas is not yet in place and consequently it is not necessary to communicate these concepts outside the company. Instead, the material created is intended to support strategic decision-making and the long-term development of the company, rather than to achieve short- or mid-term gains by creating the foundations for the next few product generations – the objective is to create motivational targets to aim at.

Product concepts for the future cannot be created without observing the signs of change in the present. This can be done by linking the product concepts to future scenarios. Future scenarios have been used as a tool to probe the possibilities of different futures in futurology[3]. The main industrial use of these studies has been to describe the possible changes in the business environment at a macro level[4]. However, these scenarios can also be used to determine the corresponding feasible product concepts. For different scenarios the company needs specific survival strategies and, of course, the company's product portfolios will also differ. Using vision concepts gives the company a new angle for analysing its responses to future challenges at a product portfolio level. A company that has made its preparations by developing scenarios and vision concepts can adjust its strategy with the aim of making the chosen vision the reality of the future. It will be able to analyse the type of product offerings that will give it the flexibility to survive and succeed in any of the various future scenarios. Implementation of the strategy may lead, for example, to new operational policies being created, new technologies being adapted, company functions being developed and new knowledge being gained or acquired from other companies.

The likelihood of the product concepts coming to fruition within the selected time frame can be evaluated using technology road maps and forecasts that are publicly available and which companies develop for their current core areas of expertise. Changes in the environment will make it necessary to update the scenarios every few years, even though the anticipated future still lies far ahead. This is important because the scenarios need to reflect the contemporary understanding of the anticipated changes in the business environment.

Influential players in their fields, such as leading companies and regulators, can steer developments to a certain extent by selecting and working towards scenarios that are favourable for them. Companies can usually identify the best and the worst alternative, which enables them to work towards the best scenario. Therefore, creating far-reaching concepts not only helps in anticipating and being prepared for changes, but also in initiating developments that will change the actual business environment.

Even though the main reason for creating vision-based concepts is to clarify long-term product strategies, this is not their only benefit. The creation of product concepts requires cross-disciplinary collaboration that includes the company's management, members of the R&D and marketing teams, together with external experts. This brings the different disciplines into multifaceted discussions about the company's opportunities and the threats it is faced with, which will very probably improve internal communications and cohesion. Working together to develop the concepts increases personal commitment to the company's long-term objectives. The process and practices of concept creation help the participants to take on board creative approaches that can bring more creativity into their daily work.

7.2   Vision concepting process

The development of vision-based concept creation methods started from an assumption that product concepts can contribute to creating a vision for a company. Concepts can be used to turn abstract ideas and objectives into something more tangible. The idea was elaborated in a cross-disciplinary research programme, TUTTI, systematic product concept generation initiative[1], whose participants included the Helsinki University of Technology, UIAH and the Institute for Advanced Management Systems Research at Åbo Akademi University. An anticipatory time span of 10 years and beyond was chosen for the project, which is the normal practice in future research. The main result of the project was a method and process description for linking scenario generation, technology road mapping and concept design for generating vision concepts. There are two main phases in the method, namely future description and conceptualisation, which are divided into thirteen steps, as described below (see Figure 7.1).

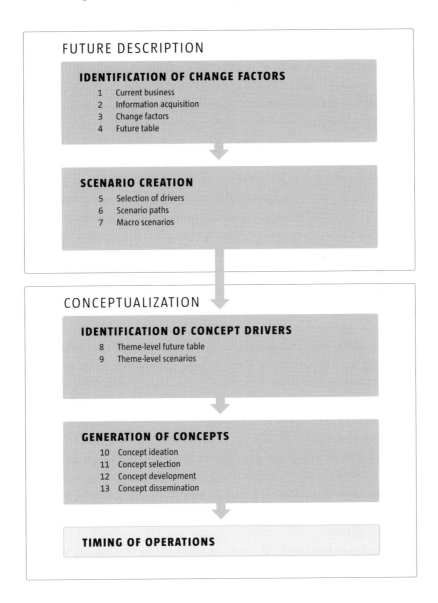

FIGURE 7.1.
The phases of the vision concept generation process

1 CURRENT BUSINESS. At the beginning of the process of creating vision concepts, the business foundations of the concepts must be identified. It is important to specify which business the company is in and which business it may be in in the future. The time span for the scenarios also needs to be decided. These issues are related, because product development cycles differ greatly between industries. For instance, several generations of new consumer electronics products can be introduced during the lifetime of a cruise ship. It is also critical to understand the core competence of the enterprise for which the concepts are being created. For instance, a company that may at first glance look like a diesel engine manufacturer can in fact be involved in the power supply business in a wider sense.

2 INFORMATION ACQUISITION. The main trends affecting the development of the company's business environment should be investigated by studying published scenarios, trend forecasts and other sources of information that anticipate future changes. For example, some well-known general and global scenarios have been developed by Shell, the CIA and the World Business Council for Sustainable Development.

The key technologies to be applied in future products should be identified. Technology foresights, technology road maps, technology studies, scientific and popular publications, and patents are all useful materials for technology anticipation. In addition, latent knowledge and information within the organisation should be collected. Technology forecasts are especially useful material for concept development, because they aim to present the expected developments in chronological order. They describe the assumed availability, development and application of technologies from the present into the future. The result is a map that illustrates how the selected technologies will develop and how and when they can be used. It is also important to monitor developments in military and space research, because advanced technologies developed in these areas are usually be available for civil purposes a few years later. Different scenarios and technology foresights have different emphases, so it is advisable to study a wide selection of them.

3 CHANGE FACTORS. In the third phase of the process, the information that has been acquired is organised using a political, economic, social, technological and environmental (PESTE) analysis. In addition to the most obvious and strongest trends, the analysis should be sensitive to weak signals that may be the first indications of major changes in the future.

4 FUTURE TABLE. The future table, also called the field anomaly realisation method[5], can be used to combine change factors (see Figure 7.2) into coherent future paths. The relevant variables identified in the PESTE analysis and their alternative values are presented in the future table by listing the variables in the rows and their different potential states in the columns. The table may include several monitoring levels, such as global, national and industry-sector levels.

The global political situation	Tense, confrontation between different blocks	Tense, everyone against everyone else	Peaceful, sensible	May cause tensions (for example, high prices in the pharmaceutical industry)	
Globalisation	continues	stops	goes backwards	slows down	
Liberalisation	continues	stops	goes backwards	slows down	
Exercise of power	local	global	decentralised		
Establishment	national political executive	non-governmental organisations	supranational companies	local political executive	decentralised
National states	status decreases	stable status	status strengthens		
International rules	everybody obeys	most of the players obey	no obedience, poor rules	everyone is forced to obey	
The leading country or block	USA	EU	Asia	China	no obvious leader
International terrorism	increases	unchanged	decreases		

FIGURE 7.2.
An extract from a future table showing political factors at a global level

5 SELECTION OF DRIVERS. Drivers are the most important possible changes that may take place in the company's operating environment. Each scenario is driven by different factors. By combining the results of company-specific PESTE analyses with existing generic scenarios, it is possible to outline the drivers for case-specific scenarios.

6 SCENARIO PATHS. A scenario-specific future path is created by systematically selecting possible scenario-specific parameter values from the future table. The creation of the path starts with the drivers, and the other values must form a logical continuum with the driver. At the end of the process, one alternative path corresponds to one alternative future scenario.

7 MACRO SCENARIOS. Scenarios are a way of summarising and documenting the results of future research. Scenarios describe alternative cross sections of the future. They should be internally consistent, plausible and logical descriptions of the potential future that include all the PESTE factors. Scenarios are typically presented as text documents that may include visual images.

8 THEME-LEVEL FUTURE TABLE. The future scenarios are relatively general descriptions of the changes at the macro level. For vision concept development, a more specific theme level has to be included in order to be able to handle more specific questions related to the product categories that are interesting for a specific company. The future table is extended by listing the future factors that affect the supply and demand of certain types of products. This means that variables describing consumers' predicted behaviour, values, lifestyle, environment and preferences concerning, for example, short-range mobility (as shown in Figure 7.3) need to be collected and placed in the future table. Political circumstances, the role and support of a society, the role of companies and individuals, class divisions, individuals' resources, assets and circumstances, the speed of technological development, possible theme-oriented boom technologies, materials and environmental questions should all be taken into account when listing the factors describing a theme.

9 THEME-LEVEL SCENARIOS. Theme-level scenarios are detailed continuations of the macro-level scenarios created earlier. These scenarios are short stories describing the combined states of the selected parameters. Again it is important that these scenarios have an internal logic, since theme-level scenarios are the basis for the actual concept design.

10 CONCEPT IDEATION. The theme scenarios are used as stimuli in team sessions where ideas for corresponding product concepts are generated. Technology road maps developed from the company's point of view are also needed, since the existing expertise that is immediately available is usually insufficient. When generating concepts for the future, we recommend focusing on questions which start with "how" and "why" rather than looking at how the product would be implemented. For example, instead of designing a futuristic lift, the focus should instead be on how people and

Need for individual space	no space	need for own space decreases	increases			
Muscle vs machine power, popularity of utility exercise	increases	no changes	decreases			
Need for movement	business travel	between buildings	inside buildings	leisure		
Social pull	bulk	disposable products	long-life	high-technology	luxury	secure
People	look for convenience	ascetic	bohemian	uncommuni-cative, scared, selfish		
Domestic animals	decrease	no changes	increase (guard dogs)			
Those who are in motion	respect tradition	adapt themselves to development	favour brand new develop-ments			
Values added to exercise	experience	pleasure	not needed	freedom	feeling of safety	
Volume	children	adults	old people	men	women	
Volume	poor	middle class	rich people			
Status value of vehicles	significant	not significant				

FIGURE 7.3.

A future table at a theme level. A sabotage scenario profile is indicated in light grey, and drivers for a short-range mobility theme are indicated in dark grey

articles will be transported in the future or even on the whole future concept of providing short-range urban mobility services.

11 CONCEPT SELECTION. Abstract ideas become concept drafts when they are given a physical form. Concept drafts should always contain some visual and textual information about the product, the user and the use context. The concept should also be clearly linked to scenario-specific features.

When creating design variations a single concept draft may require a large number of sketches and ideas before the correct configuration is found. Ideation sessions can generate a vast amount of ideas that are widely distributed around the problem area, even though restrictions and specifications have been put in place. The time span and the level of realism of the ideas also vary widely. Concept ranking, grouping, rejection, combination and selection procedures are needed, for which there are a range of methods available, such as concept scoring and concept screening[6]. External decision-making, product championing, intuition and multiple voting may also be used. Support in the form of technological expertise must be provided when imagining how future products could work and be realised. However, in vision concept development the technology may not necessarily exist yet.

The business-positioning field (Figure 7.4) can be used to illustrate the link between vision concept choices and business strategies. Concepts can be placed in a grid where the vertical axis represents the technological change, technical knowledge and expertise, and the horizontal axis represents market development and knowledge about customers and markets. The further away you move from the origin, the more you need to acquire knowledge and develop your operations. If the company wants to stay within its existing technical competences but expand into new markets, it must choose the concepts in the lower right-hand corner. If it seems more appropriate to remain in familiar markets and manufacture products that need new technologies, the management should choose the product concepts in the upper left-hand corner.

FIGURE 7.4.
The business-positioning field

12 CONCEPT DEVELOPMENT. The concept drafts with the most potential are developed further. Concept-specific technology studies and market-potential assessments should be carried out to enable further and more focused concept development.

13 CONCEPT DISSEMINATION. Final designs, technology and market specifications are created, and the concepts are presented using visualisations, mock-ups, miniatures or animations. Concepts developed for strategic purposes must contain sufficient information about the concept idea. The exact appearance or structure is not very important because it will change as a result of the long time frame, but will play a more significant role if the concepts are used in marketing or advertising that is presented to the public.

7.3 Six global scenarios

It is impossible for a company to prepare for the future by defining only one scenario, because economic, technological and social systems are so complex and unpredictable. The aim of future anticipation is therefore to identify several alternative future scenarios as accurately as possible. The main issue is not whether or not the predictions turn out to be correct, but more importantly how well the set of scenarios enables the company to map the alternative futures[7].

In the TUTTI project, scenarios were created from the perspective of the Finnish technology industry. The driving forces for the scenarios were selected from a list of identified change factors that had been classified using a PESTE analysis. They were derived from technological, market and societal dimensions, and the six final scenarios illustrate coherent development in all of these areas[1,8]: tightening competition scenario, emerging markets scenario, Europe-centred scenario, sabotage scenario, changing values scenario and biotechnology boom scenario.

The *tightening competition* scenario is a "business as usual" case that extrapolates many of the present trends into the future and indicates a world with a plentiful supply of goods, but with tightening competition and increasing efficiency and profit requirements. The USA is the only super-power, but the focus of the global economy may be gradually shifting towards Asia. However, traditional business practices and consumer behaviour patterns continue to dominate.

In the *emerging markets* scenario, the focus of the global economy, the markets, economic growth and volume shifts to Asia. Although global integration continues, there will be several new players in the global economy. China is on its way to becoming the largest economy in the world[9,10] and Asian business practices may come to dominate the global economy. This scenario highlights the challenges presented by new, emerging markets and their cultural characteristics.

In the *Europe-centred* scenario, European countries acquire more influence in international organisations and forums, such as the UN. Russia and the EU may become more closely integrated, which may make considerable oil, gas and raw material resources available to European industry. In addition, the EU improves its position in technology and science. Basically, this scenario describes the new blossoming of Europe and its consequences for the global economy.

In the *sabotage* scenario global development is characterised by discontinuities, such as local and global conflicts, terrorism, environmental disasters and a series of epidemics. A growing lack of communication, restrictions and insecurity has a major influence on the economy. This scenario highlights possible future threats that could jeopardise the overall positive development of the global economy.

In the *changing values* scenario, consumers' values change dramatically. The main issue is the power shift from multinational corporations to non-governmental organisations. Consequently, local brands will be more successful than global ones. This scenario may lead to a "no logo" development[11] or even to the growth of extreme alternative environmental movements. This scenario illustrates the rise of the post-modern anti-materialistic values that prioritise cultural and environmental issues over economic ones.[12]

In the *biotechnology boom* scenario, revolutionary developments in biotechnology change the world. Biotechnology supplements or even replaces other technologies. Several new technologies emerge, and biotechnology achieves full consumer approval. The consequences for society, the markets and other technologies would undoubtedly be significant.

These six global scenarios explore alternative future states in which industries and companies have to be able to operate, and illustrate possible changes in technical development, global and regional markets and societal development in different geographic areas. The scenarios are, of course, only

FIGURE 7.5.
The step-in car concept derived for the sabotage scenario

exaggerated examples of the possible future outcomes, but they are considered relevant. After the global scenarios have been created, they need to be extended into the business, theme and product levels in order to examine what kinds of opportunities and threats would emerge.

7.4   Concepts for short-range mobility and robotics

In this section we look at four examples of vision concepts. Two of them, the *step-in car* and the *mall mule*, are related to short-range mobility, and have been developed for the sabotage and tightening competition scenarios, respectively. With these concepts the technology platforms change, but the business remains the same. The second two vision concepts, the *fly-tron* and the *automated textile machine*, illustrate the possibilities of robotics applications, and have been designed for the sabotage and changing values scenarios, respectively. Here the technology platform remains the same, but the market changes.

The step-in car (Figure 7.5) is a two-passenger light vehicle that fulfils the need for safe, urban transportation in the sabotage scenario. It makes short-distance travel possible without requiring the passengers to step outside, and hence protects its passengers from external threats such as pollution, noise and violence. In addition to the actual protection, the step-in car offers the feeling of safety, privacy and integrity. A high level of safety is achieved by the use of advanced materials in the chassis construction. The car must be able to use several alternative fuels in order not to be subject to problems in fuel distribution.

The mall mule (Figure 7.6) is an automatic transport system for people in department stores and malls. The mall mule will handle transportation inside the huge malls of the tightening competition scenario, as well as

FIGURE 7.6.
The mall mule concept developed for the tightening competition scenario

FIGURE 7.7.
The fly-tron concept for the sabotage scenario

transport between the parking areas and the malls. The mall mule enables department stores and malls to control and guide the flow of customers and, with the cart's multimedia connection, customised services such as advertisements, notifications, product presentations and payment methods can be provided for the passengers. The concept consists of an automatic transportation system for people, passage control and an enclosed service system. The system is based on guided vehicle technology in which the cart constantly monitors its environment using cameras and sensors. The carts are charged and stored in compact stacks and special storage towers.

The fly-tron (Figure 7.7) is an insect-like surveillance gadget equipped with several sensors. Indoors, for example at home or in hotels, the fly-tron ensures a safe environment by autonomously monitoring the quality of the air, the integrity of property, the health of the owner and other safety and security-related issues. The concept responds to the increasing personal safety and environmental surveillance needs that arise from the sabotage scenario. In order to implement the fly-tron, lightweight energy storage techniques and lighter, more sensitive cameras and sensors that use less energy must be developed. Contour identification systems and solutions for autonomous movement must also be developed further. In spite of the challenges, no technological revolution is needed.

The *automated textile machine* (Figure 7.8) is an automatic manufacturing unit for clothes for use at "clothing cafeterias". The concept responds to brand adversity and the prioritising of local production in the changing values scenario. The automated textile machine is a miniature clothing factory comprising manufacturing machinery, a customer interface and a 3D scanner with which the machine records the customer's personal measurements. The customers are able to produce garments according to their preferences by choosing and combining models and fabrics from different design catalogues or by creating a model themselves. The garment is then produced automatically.

During the creation of the concept examples it became clear that often it is not sufficient to concentrate on an individual product – the role of an individual product needs to be seen as a part of a system. As an example, it is difficult to imagine the forklift truck of the future without thinking about the system around the product: where are the forklifts used, what kind of transport facilities are available, how have other vehicles and products developed, what are the possible standards and what kind of cargo-handling

FIGURE 7.8.
The automated textile machine for the changing values scenario

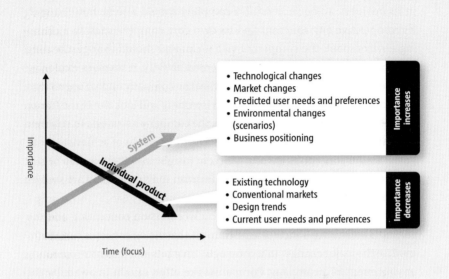

- Technological changes
- Market changes
- Predicted user needs and preferences
- Environmental changes (scenarios)
- Business positioning

Importance increases

- Existing technology
- Conventional markets
- Design trends
- Current user needs and preferences

Importance decreases

Importance

System

Individual product

Time (focus)

FIGURE 7.9.

Characteristics of product-oriented vs system-oriented concept development

infrastructure is in place? In product development for the near future we are familiar with the target markets and the ways of applying existing technologies. It is also possible to collect reliable user information, and the role of the design and the brand are important. When the focus is shifted to the future, it becomes important to find out how the technologies and markets are likely to change, to identify the surroundings and to predict future user needs (Figure 7.9). For example, in the TUTTI project we realised that some of our concepts offered features and solutions that were too detailed, and which therefore focused attention on product design rather than stimulating discussion about systems, infrastructures and opportunities on a more general level.

7.5   Conclusions

Every company has a vision, which is the objective it is trying to achieve in its business. To be successful, a company should also be familiar with its competitive environment and its own core competencies. In a changing environment, the company's vision can and should change. Creating a vision is, by definition, a future-oriented activity. It involves exploring possibilities and trying to anticipate what the company and its operations, core competencies and competitive environment will look like in the future. The tool for implementing the vision is the company's strategy. In strategic planning the company defines the goals and the means of achieving those goals. Vision concepting gives a concrete insight into the possible future product portfolios and links strategic decision-making with future studies and concept development.

One of the biggest differences between vision concepting and the concepts created for a shorter time frame is that vision concepts can accommodate the major changes in technologies, markets and the entire operating environment of a company. Companies are often caught unaware by the gaps between old and new technologies; this problem can be avoided by carrying out visionary projects. Vision concept development helps companies to benefit from being the first to use a new technology or to develop a new market. However, it is important to check the scenarios systematically because the business environment can change rapidly.

Vision concept development involves multidisciplinary teamwork where the team members come from a wide range of corporate functions and also from outside the company. The team members should include top management executives, product development engineers, sales and marketing people, industrial designers and futurologists. Creative work is stimulating, and produces fascinating results that affect daily routines and current product development. Concept development with a long time frame gives an excellent opportunity for cooperating with other companies, even competitors. This is a current practice in the telecommunications business, for instance, which could also be adopted by other industries. The important factor is that the different players who participate in the development of a specific system may have additional common initiatives and aims.

Concept creation helps to improve companies' understanding of the markets and the environment. This understanding can also contribute to the development of better-targeted products. In addition when the futuristic concepts are presented to the public for marketing purposes, they create an image of a company that is a genuine pioneer in its field.

7.6 References

1   **Kokkonen, V., Kuuva, M., Leppimäki, S., Lähteinen, V., Meristö, T., Piira, S.and Sääskilahti, M.** (2005). Visioiva tuotekonseptointi – Työkalu tutkimus- ja kehitystoiminan ohjaamiseen [Vision concepting – An approach for R&D management]. Teknologiateollisuus, Helsinki.

2   **McDermott, C. and O'Connor, G.**, "Managing Radical Innovation: an overview of emergent strategy issues": The Journal of Product Innovation Management, volume 19, 2002, 424–438.

3   **Schwartz, P.**, "The Art of the Long View: Planning for the Future in an Uncertain World": Currency, 1996.

4   **Schoemaker, P.**, "Scenario planning: a tool for strategic thinking": Sloan Management Review, 1995, 25-40.

5   **Rhyne R.**, "Projecting Future Contextual Scenarios", The FAR Method Revisited, (http://www.risk-sharing.net/ceci/Resource/Rhyne_Russell/, 17th March 2005), 1998.

6   **Ulrich, K., and Eppinger, S.**, "Product Design and Development": McGraw Hill, USA, 2000.

7   **Matthews, W.**, "Kissing Technological Frogs: Managing Technology as a Strategic Resource": European Management Journal, volume I 9 (2), 1991.

8   **Sääskilahti, M., Kuuva, M., Leppimäki, S.** A Method for Systematic Future Product Concept Generation. International Conference on Engineering Design, Melbourne 2005, paper submitted.

9   **China in the World Economy: the Domestic Policy Challenge, Synthesis report, OECD, 2002.**

10 **Chow, G.**, "China's Economic Transformation": Blackwell Publishers Inc., USA, 2002.

11 **Klein, N.** (2000) No Logo – Taking Aim at the Brand Bullies. Picador USA.

12 **Inglehart, R.**, "Globalization and Postmodern Values", Washington Quarterly (Winter 2000), 23:1, 2000, 215-228.

Contributor Biographies

Turkka Keinonen

Turkka Keinonen works as a professor of industrial design at University of Art and Design Helsinki. His teaching and research interests lie in user-centered industrial design and product concept creation. Keinonen currently acts as the head of research at the School of Design at the University of Art and Design Helsinki leading a unit of 20 researchers. He is in charge of several research projects focusing on method development for user research and for evaluating and organising design in the information and communication technology and engineering industries.

Prior to his current position, Keinonen worked for several Finnish design consultancies and in the usability group of Nokia research center. In his doctoral thesis, Professor Keinonen studied the influence of usability on consumers' product preferences.

Keinonen has written or co-edited the following books: *One-dimensional Usability* (1998), *How to design for usability?* (2000, in Finnish), *Mobile Usability – How Nokia Changed the Face of the Mobile Phone?* (2003), *Product Concept Design* (2004, in Finnish) and *The Change of Industrial Design* (2004, in Finnish). He is a co-inventor in 25 international patents. He was nominated the designer of the year 2000 in Finland for his work in user-centered industrial design, and in 2003 he received The Federation of Finnish Electrical and Electronics Industry SET prize for developing usability education.

Turkka Keinonen
University of Art and Design Helsinki
Hämeentie 135 C
00560 Helsinki
Finland
turkka.keinonen@uiah.fi

Roope Takala

Roope Takala is a program manager at the multimedia technologies laboratory of the Nokia research center. His current work involves managing research and technology implementation in the domain of human–machine interaction. His prior work includes the management of projects in active haptics, electromechanics integration, competitor benchmarking and developing system for product concept demonstration. Takala has also worked as a visiting scholar in the center for innovation in product development at the Massachusetts Institute of Technology. His work concentrated on product concept creation and evaluation-related research. Takala has had several journal and conference papers published on product concepting, environmentally sustainable engineering design and design for assembly. He currently has two patents, and has three patents pending.

Roope Takala
Nokia Research Center
Itämerenkatu 11–13
00180 Helsinki
Finland
roope.takala@nokia.com

Vesa Jääskö

Vesa Jääskö currently works as a design strategist at the design consultancy Muotohiomo, where he provides design expertise to activities in preparation of product development projects and in conceptual design. Jääskö also works as an industrial designer. Prior to his current position Jääskö worked as a researcher at The University of Art and Design Helsinki and as a professor at the University of Lapland.

Vesa Jääskö
Muotohiomo
Lapinlahdenkatu 31 B
00180 Helsinki
Finland
vesa@muotohiomo.com

Jussi Mantere

Jussi Mantere received his MSc in usability and user-centered design from the Helsinki University of Technology. Currently Mantere is working at the Nokia design unit responsible for developing methods for the evaluation and validation of product concept design, he has previously held positions at Nokia related to user experience, user-centered product development processes and the usability evaluation of products.

Jussi Mantere
Nokia
Keilalahdentie 2-4
02150 ESPOO
Finland
jussi.mantere@nokia.com

Toni-Matti Karjalainen

Toni-Matti Karjalainen, who holds a DA in art and design, works as a project manager and researcher at the Helsinki University of Technology and at the University of Art and Design Helsinki. His main research topics include product design, product semantics, brand management and product development. The research activities of Karjalainen are closely connected with both Finnish and international companies. Karjalainen has published numerous articles in academic publications and the popular press. He also teaches and lectures in various courses, seminars and conferences in different countries.

Toni-Matti Karjalainen
Helsinki University of Technology
BIT Research Center
P.O. Box 5500
02015 Espoo
Finland
toni.karjalainen@uiah.fi

Anssi Tuulenmäki

Anssi Tuulenmäki works as a project manager at the Helsinki University of Technology, in the BIT research centre. His multidisciplinary team (decode. hut.fi) researches product development and design in cooperation with companies and institutional investors. Tuulenmäki is especially interested in the creation of new market spaces, and lectures at companies, universities and international conferences.

Anssi Tuulenmäki
Helsinki University of Technology
BIT Research Center
P.O. Box 5500
02015 Espoo
Finland
anssi.tuulenmaki@hut.fi

Mikko Sääskilahti

Mikko Sääskilahti works as an industrial designer for SWECO PIC, an international product development and engineering services consultancy. Sääskilahti has worked in boat, forestry machinery, elevator and numerous consumer product design projects. Previously Sääskilahti worked as a researcher at the Helsinki University of Technology in machine design and in the TEKES project on visioning concept development. He is currently pursuing a PhD degree at the Helsinki University of Technology .

Mikko Sääskilahti
SWECO PIC
P.O. Box 31
01601 Vantaa
Finland
mikko.saaskilahti@sweco.fi

Subject Index